Introduction to the Study of the Neural Basis of Action and Thought

Authored By

Pierre Nelson and Gérard Lot

40-46, rue de la Renardière
93100-Montreuil
France

General:

1. Any dispute or claim arising out of or in connection with this License Agreement or the Work (including non-contractual disputes or claims) will be governed by and construed in accordance with the laws of the U.A.E. as applied in the Emirate of Dubai. Each party agrees that the courts of the Emirate of Dubai shall have exclusive jurisdiction to settle any dispute or claim arising out of or in connection with this License Agreement or the Work (including non-contractual disputes or claims).

2. Your rights under this License Agreement will automatically terminate without notice and without the need for a court order if at any point you breach any terms of this License Agreement. In no event will any delay or failure by Bentham Science Publishers in enforcing your compliance with this License Agreement constitute a waiver of any of its rights.

3. You acknowledge that you have read this License Agreement, and agree to be bound by its terms and conditions. To the extent that any other terms and conditions presented on any website of Bentham Science Publishers conflict with, or are inconsistent with, the terms and conditions set out in this License Agreement, you acknowledge that the terms and conditions set out in this License Agreement shall prevail.

Bentham Science Publishers Ltd.
Executive Suite Y - 2
PO Box 7917, Saif Zone
Sharjah, U.A.E.
subscriptions@benthamscience.org

BENTHAM
SCIENCE
PUBLISHERS LTD.

CONTENTS

Foreword

As a physician specialist in urodynamics (and a researcher in mathematical modeling of lower urinary tract and micturition), I need to connect the complaints of my patients to physiological causes. In most cases causes can be myogenic, neurogenic or both. Nervous excitations are poorly defined and poorly measured (because very invasive) in human. We know that the brain plays a great part in the control of micturition. But cognitive sciences do not teach us anything in connection with the conscious aspect of control. From functional imagery, we learn that some changes in the map of activated areas come with some voiding dysfunctions. However all this is awfully fuzzy.

From non-invasive to invasive, proposed treatments of voiding dysfunction are rehabilitation, chemical such as α-blockers or anti-muscarinic, sacral neuromodulation, botulinum toxin and at last surgery.

Rehabilitation probably induces new behaviors and therefore new reflexes while chemical treatment modifies intercellular transmission and neuromodulation or botulinum toxin the synaptic transmission.

Many studies are carried out on a small step on one phenomenon and unfortunately on animals. Thus I felt strongly the need of a synthesis.

Then, I met two physicists: Pierre Nelson and Gérard Lot. After a bright career, they devoted their mastery of mathematical modeling applied to practical problems to the study of nervous activity. Proceeding by little steps, they started from the physicochemical properties of a neuron to obtain input-output relationships, which are the description of its processing signals ability. From then, they could overlook all physicochemical properties and look only at the functional ones. If large assemblies of parallel, competitive or modulating neurons explain easily the behavior of frog, human mechanisms are much more intricate and their description complicated. However, I think that the effort of description and modeling of these specialists is a great step in the understanding of action and thought.

Françoise A. Valentini
Université Pierre et Marie CURIE (Paris 06)
Physical Medicine and Rehabilitation Department
Rothschild Hospital
Paris
France

Preface: The Goal and the Method

Our purpose is to describe how the activity of physiological neurons (and not schematic ones) generates behaviors and thoughts of animals and men. We own a huge number of experimental results (the number of notable recent papers is nearer of 10^5 than of 10^4).

Systems such as the brain or the sun are so complicated that we need a hierarchy of models to study them at several description levels. A model is a quantitative (mathematical) description summing up a large number of experimental observations. It establishes relations between observable parameters. Quantitative modeling is the convenient tool for a bottom-up approach. Nervous system can be looked at from very different points of view: nature of the specific molecules, electric phenomena, anatomy, circuitry, behavior, psychology. For each of these levels, we own a huge number of scattered experimental results. We intend to start from the lower level, then to proceed to its accurate study, and finally to extract from this description some results simple enough to allow the study of a higher level.

The nervous system of any animal links input sensorial cells (or sensors) to output cells (muscles and glands). A network of excitable cells, the neurons, carries out the connection.

From the most primitive animals to the man, we do not perceive any significant evolution in the basic component: the neuron. The increase in the sophistication of the nervous systems is due to an increase in the number of neurons and not in an improvement of the performance characteristics of the neuron itself. Some sea anemones count only one neuron that simultaneously acts as a sensor and as a motoneuron: when directly excited, it orders tentacles closure. Instead the nervous system of human being is build around some 10^{11} neurons, half of them devoted to basic functions such as sensation and motorization, the others to more elaborate functions such as memorizing, speaking and thinking. Starting from the general properties of the nervous components, we will go gradually towards the observable behavioral of various animals from the simplest ones to the man.

Analogies with Computers

The neuron is not a chip, the brain is neither a computer nor a cellular automat or an artificial neuronal network, but both brain and computers are messages processing devices. Some similar laws, many of them discovered by engineers (for instance the part of noise or the difference between quick and slow memories) rule them. Thus, analogies can be useful.

A Text Written to be Easy to Read:

Although using many theoretical and computational results, this text, divided in 5 chapters, can be read without any mathematical background: the main text contains only a qualitative description of the main assumptions and the main results. Quantitative results and descriptions of the used theoretical methods are given in appendices at the end of each Section. We advise the reader to look first only at the main text, then to the quantitative results of some appendices and only then to the complete demonstration (*in italics*) of the most difficult questions.

To gather an exhaustive bibliography would be an impossible task. Thus, only some very important or seldom read papers are referenced in this text.

A Study Without Any Philosophical Involvement

We assume that the brain is made of molecules (and this is the contrary of spiritualism), but that its working requires a marvelously structured architecture (and this is the contrary of brute materialism or of reductionism). We pretend that the basic circuitry is the conditional link (and this is close to behaviorism) but that the drives, the rewards, the wirings are, in most cases, very different from those encountered in the simplest pavlovian reflex (and this is the contrary of behaviorism). To sum up, we believe that the neuro-physiological phenomena are far too complex to be confined in black or white doctrinal concepts.

ACKNOWLEDGEMENTS

The authors are deeply grateful to Dr. Françoise Valentini who undertook the exhausting task to correct the presentation of this text and to Dr. Humaira Hashmi and Hira Aftab who patiently advised them.

CONFLICT OF INTEREST

The author confirms that this eBook contents have no conflict of interest.

Pierre Nelson and Gérard Lot
40-46, rue de la Renardière
93100-Montreuil
France

CHAPTER 1

Neurons: From Ions and Molecules to Messages Transformations

Abstract: The excitability of neurons is due to electrical and chemical phenomena. Motion of ions is described by currents and potential, moreover intricate chemical cycles are described by amplitudes and delays. Starting from what is usually known about synapses and Hodgkin-Huxley axon, we describe first all the parts of a standard neuron (from its input till the input of the following neuron). Then, looking at various physical constraints, we generalize to any kind of neuron. Thus, the synthesis of the physicochemical properties of the cell enables us to compute when spikes occur. Now, we show that the firing rate is the significant nervous message. So, from the input – output relations, we are able to compute the processing abilities of any neuron (some of them make linear additions; others exhibit an ON-OFF behavior, and so on). The next step is to evaluate limitations caused by noise and to study little sets of neurons (of simple animals or very well localized in man, for instance the center controlling breathing). Then, we begin to look the three main functions of huge neuronal sets: to code sensorial message, to choose between competitive signals, to modulate other neuronal sets.

Keywords: Axon, channels, control potential, elementary operations, generalized Ohm's law, G protein, heart oscillators, Hodgkin-Huxley axon, hormones, ionic concentrations, modulations, Nernst potentials, pheromones, pneumotaxic center, rest potential, spike, synapse, threshold, transmitter, typical neurons.

1.1. What Everybody Knows of Neurons

1.1.1. The Sequence of Events

Aristotle confounded nerves with tendons and ligaments. We know today that the nervous system of man and of most of animals is made of a brain and from the *afferent* nerves going from the organs towards the brain and the *efferent* nerves going from the brain towards the organs, most of the nerves following the spinal cord.

With a microscope, we look at a slice of nervous matter stained by silver salts. We see clearly the nervous cells (the neurons) wrapped by unexcitable cells, the glial cells, which support and feed the neurons, but do not play any part in our study.

Neurons (Fig. **1.1**) exhibit a large variety of sizes and shapes [1].

Pierre Nelson & Gérard Lot

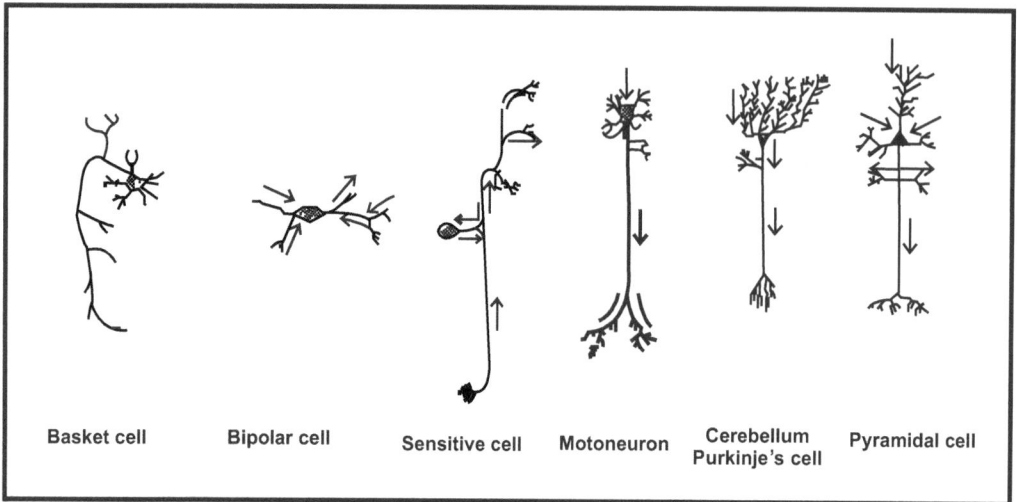

Fig. (1.1). Shape of some neurons (Arrows show the direction of propagation of the signal).

But all of them comprise the same elements (Fig. **1.2**): a cell body containing the nucleus, the dendrites supporting the synaptic inputs, an axon and a terminal arborescence supporting the synaptic outputs. The axon can be very short or very long (about 1 m). Long axons are covered by an isolating white matter (the myelin) with some nude places (the Ranvier's nodes). Nervous matter without myelin is grey.

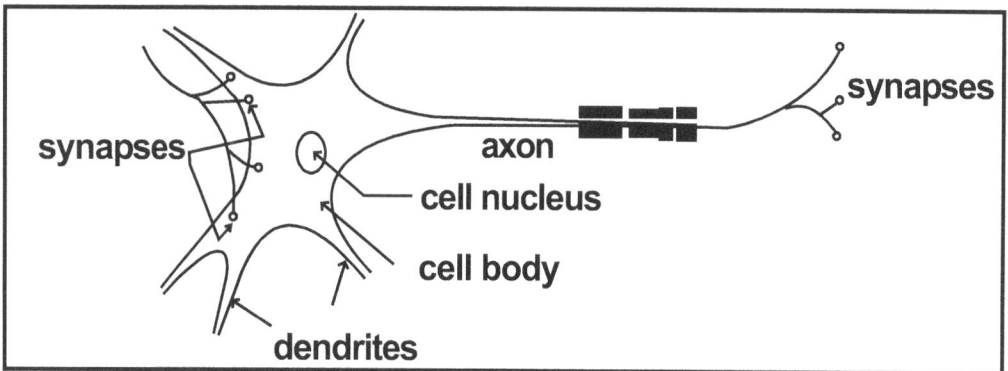

Fig. (1.2). Components of a neuron.

We introduce an electrode into the axon of a squid neuron (chosen because it has a big diameter) and we connect it to a scope. Thus, we measure the electric potential of the inside of the cell. When the neuron is silent, we find a constant *rest potential* V_R = -70 mV. We verify that it does not vary along the axon. When the neuron is excited *in vivo* by other neurons (which creates a *control potential* slightly greater than V_R, for instance V_C = -60 mV), we observe either salvos of regularly spaced electric *spikes* or a sequences of irregularly spaced ones. All the spikes are alike.

Their width is less than 1 ms, their top is +47 mV. After a spike, we observe oscillations of the potential. The bottom of the oscillation is –80 mV. The spikes firing rate (the number of spikes by second) is seldom greater than 300 s^{-1}. This firing rate increases with the control potential.

Spikes propagate along the axon without shape change and with a velocity depending only on the axon diameter. Then, it reaches the outputs of the neuron: the synaptic buttons.

Look now with an electronic microscope. We observe the connection between two neurons: a *synapse* (Fig. **1.3**). A synaptic space separates the first neuron from the second one. The end of the first neuron is a synaptic button. *Transmitter* molecules are synthesized by the specialized cell organites (the mitochondria) and stocked in easily observed vesicles (Fig. **1.3** hereafter).

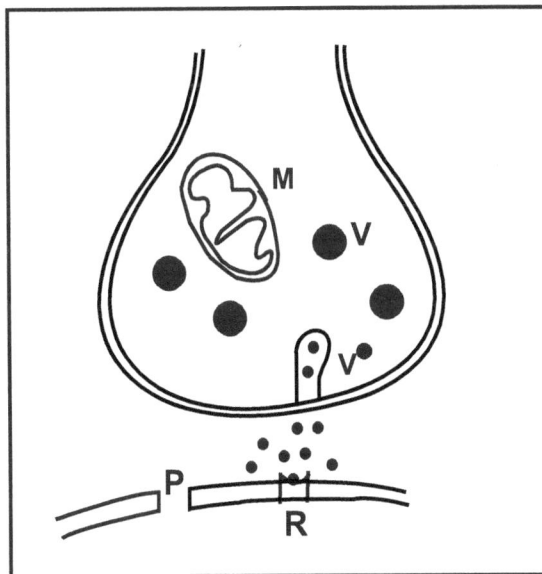

Fig. (1.3). Connection between two neurons. *M: mitochondria – V, V*: closed or open vesicle – R: receptor – P: post-synaptic channel.*

When a spike reaches the synaptic button, vesicles move towards the membrane and release transmitter molecules into the synaptic space. They diffuse in this space and later out of it; they are destroyed or recaptured by specialized molecules. The beginning of the second neuron is a membrane containing *receptors* and *post synaptic channels* (also called post *synaptic pores*). Receptor binds reversibly and selectively with transmitter molecules. When activated, it opens (directly or by releasing some *messenger*) the gate of the channel.

Ionic flow through an open channel creates a local *micro-potential*. The micro-potentials of all the synapses add in a complicated way to generate a control potential (which we had mimicked with our second electrode).

The Fig. (**1.4**) sums up this complex sequence of events.

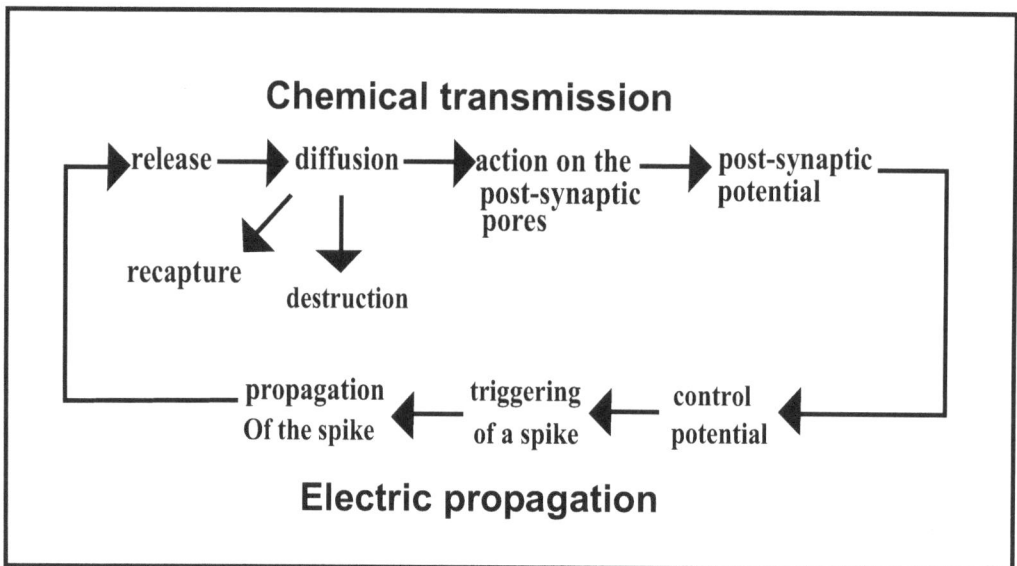

Fig. (1.4). The sequence of events.

1.1.2. The Electric Phenomena

Main fact: The control potential induces the firing of the neuron.

1.1.2.1. Ionic Concentrations and Nernst Potentials

Since the discovery of electricity and Galvani's works, one knows that nervous activity is linked to electric activity (that is to electric potentials and currents).

More recently, the neuronal media were analyzed. A blood barrier separates the outside medium of a nervous cell (the cerebrospinal fluid) from the blood. (We will see below that one effect of this barrier is to multiply the spikes firing rate by a factor of about 100). The inside medium of any cell is separated from the outside by the cell membrane. Both inside and outside media are ionic solutions in water. Their detailed balances differ: the outside medium contains mainly sodium and chlorine, the inside medium potassium, proteins and carbonic acid. (Appendix A). These concentrations are strictly controlled.

1.1.2.1.1. The Making Up of Concentration Gradients

In neurons as in any cell, the respiratory chain products a cellular fuel, the ATP. In neurons (Appendix B), the main part of this fuel is used to continuously pump out sodium and pump in potassium ions with a very high energetic efficiency [2]. Other kinds of pumps allow maintaining constant the various gradients of ionic concentrations.

1.1.2.1.2. Nernst-Donovan potentials

Pumps create concentrations differences between the inside and the outside. Neurons use this difference to create fierce electric phenomena. Imagine a hole between the inside and the outside compartments. Ions flow from the high concentration medium toward the low concentration one. Flows of charged particles are electric currents. So, we observe a Na^+ current, a K^+ current and so on. The total electric current is the sum of all the ionic currents. The same electric potential V acts on all the ionic species.

Each ionic species behaves as an electric battery: it is governed by an electromotive potential called Nernst-Donovan's potential and by an internal resistance (which does not obey to the usual linear Ohm's law; see Appendix C). The main fact is that each ionic species owns a Nernst potential associated with a zero ionic current and which depends only of the ratio of the inside and the outside concentrations (Appendix C).

For the standard concentrations, the Nernst potentials of the ions Na^+, K^+ and Cl^- are 47, -80 and -67mV.

1.1.2.2. One Particular Part of a Particular Neuron: the Hodgkin-Huxley (HH) Axon [3].

The giant axon of the squid Loligo was the first to be studied because it is a homogeneous cylinder (a very simple geometry) with a great diameter (allowing easy use of electrodes). When a constant electric excitation is applied to one end of the axon, electrodes placed along the axon detect a sequence of electric spikes which propagate toward the second end.

1.1.2.2.1. The Shape of the Spike is Closely Related to the Nernst Potentials

We look at the numerical values of characteristic points of the potential *versus* time curve during a spike (Fig. **1.5** below). The rest potential V_R is between the Nernst potentials for K^+ and Cl^-. The top of the spike is equal to the Nernst potential for Na^+. The bottom of the curve is equal to the Nernst potential for K^+.

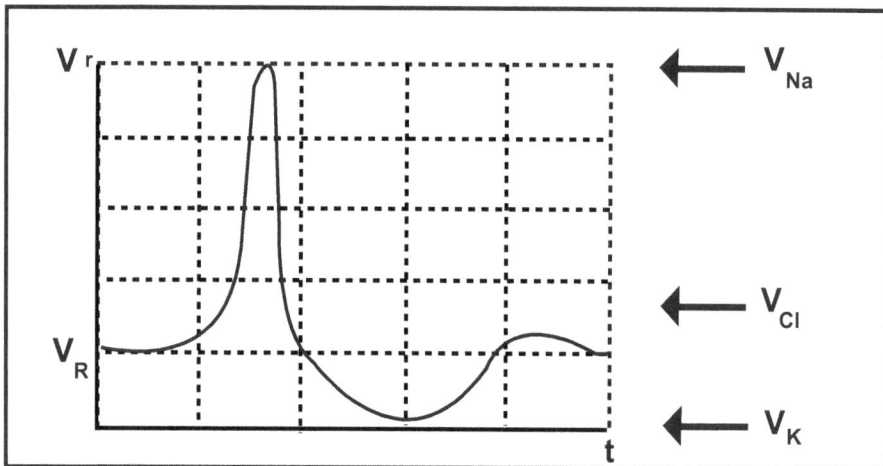

Fig. (1.5). Neuron potential is related to the Nernst potentials.

1.1.2.2.2. Qualitative Interpretation

The rest potential is due to a compromise between weak K^+ and Cl^- leakages. Excitation of the axon causes at first a very strong leakage of only Na^+. The inside potential tends toward the Na^+ Nernst's potential. Then, the Na^+ leakages stop and K^+ leakages begin. This process implies that the cell membrane is not drilled by simple holes, but by specialized channels exhibiting two strange properties: selectivity (Na^+ and K^+ ions flow through different channels) and voltage gating (the channels can be open or closed; the ratio Ω of open channels of each kind depends in a complicated and delayed way of the electric potential inside the axon).

Channels have been isolated, cloned and their molecular structure has been studied [4]. They are assemblies of proteins crossing several times the membrane and grouped around a pore. A poorly studied gating loop locks the pore.

1.1.2.2.3. The Mechanism of Selectivity

Sodium, potassium, chlorine and calcium channels are known. The radius of the pore, the electric polarization and the chemical affinity of the pore wall explain selectivity (Appendix D). This information will not be useful for our purpose.

1.1.2.2.4. Voltage Gating

The patch clamp method [5] allows to measure *in vitro* the current *versus* potential relation of one isolated channel. Each channel can be obstructed by several gates. A channel is open if all its gates are open. When the control potential is sufficient, the Na^+ channel opens quickly, then closes slowly (about 2 ms) (Fig. **1.6**) while the K^+ channel opens slowly. Cl^- channels are always open.

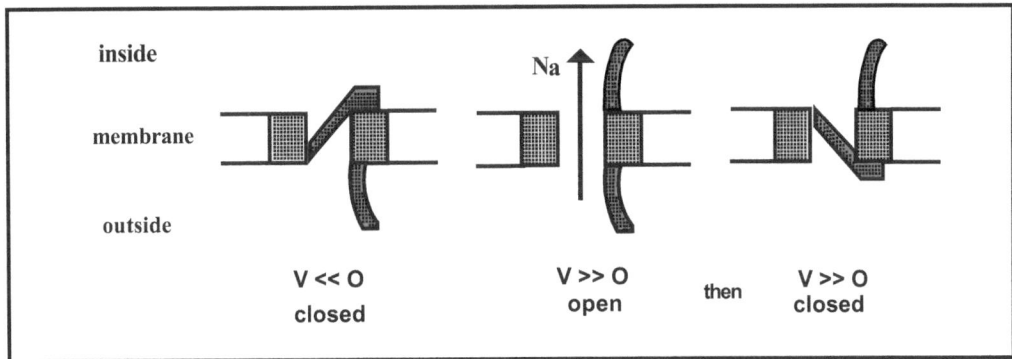

Fig. (1.6). Opening and closure of a Na^+ channel.

When applied to the several strands of the gating loop, the simplest chemical kinetics leads to the observed laws (Appendix E). This discussion will be used to understand how heart excitable cells work. The theory introduces 6 functions of the potential V (four of them for the Na^+ channels and two for the K^+ channels):

the equilibrium values $m_{eq}(V)$, $n_{eq}(V)$, $h_{eq}(V)$, and the time constant $\tau_M(V)$, $\tau_N(V)$ and $\tau_H(V)$.

1.1.2.2.5. A Synthesis: The Hodgkin-Huxley's Axon

We look at a homogeneous cylinder with radius R. Its membrane is a lipidic structure with a dielectric constant $\varepsilon = 6$. Its thickness is about 5 nm, its electric capacity is $C = 1$ mF/cm^2. Channels and pumps are embedded in it. An axial ionic current flows inside; axial current and potential are related by the usual Ohm's law. Lateral ionic currents (leakages) flow through the selective channels. They are governed by a generalised Ohm's law. The global equations are established in Appendix F. They are somewhat similar to the equations describing the propagation of an electric signal along a coaxial cable. Their solutions exhibit the creation and the propagation along the axon of a sequence of spikes.

1.1.2.3. Calcium Ion as a Cellular Messenger

Ca^{2+} pumps and mitochondria bring back the inside calcium concentration to zero in a neuron at rest. Thus, the entrance of Ca^{2+} during each spike, although very weak, is a very sensitive signal: free Ca^{2+} acts as a "second messenger" that signals to the organites inside the cell that the cell has been stimulated. Calcium governs transmitter release at the end of neurons, Na^+/K^+ pumps control, several self-regulating cellular loops, long term memory and contraction of muscular fibres.

Calcium channels are imperfectly selective. They are gated with a very weak delay τ. From the generalized Ohm'law (Appendix C) and the gating law (Appendix G), we deduce that Ca^{2+} enters into the neuron if -25 mV $< V < +25$ mV (that is, only during a spike).

1.1.3. The Synaptic Transmission

Main fact: The difference between swift and slow synapses.

An accurate knowledge of the molecular phenomena in synapses led to tremendous pharmaceutical progress. So, every detail had been carefully studied in a crushing mass of publications. Our goal is different: we only want to establish the relation between the firing of a first neuron and the opening of the post-synaptic channels of the following neuron. In this survey, we report the minimum of facts one has to know to reach this objective. Thus, we will neither describe yet the logistic phenomena (synthesis, transport, repairing of the different molecules) nor the electric gap junctions between two neurons although they play a part in the working of human retina (see Section **3.1**).

1.1.3.1. Common and Uncommon Synapses

At first view, a common synaptic link is very similar to a connection between two electric components: the link is local and selective. But, we know also uncommon synapses:

Modulating synapses are not selective. The synaptic button is connected to the post synaptic areas of all the neurons of a small volume. The synapse modulates all of them.

Hormone releasing synapses of the (hypothalamus + pituitary gland) system do not act locally. Hypothalamus, a little set ($\approx$ 5 g) of some 10^8 neurons releases two peptidic hormones (vasopressin and oxytocin) in the general blood flow and several releasing factors (for instance the corticotrophin releasing factor CRF) in the pituitary portal vessels which are the link between the hypothalamus and the nearby pituitary gland (Fig. **1.7**). Pituitary gland is very small ($\approx$ 0.5 g). When excited by releasing factors, it releases hormones, for instance the adrenocorticotrophin ACTH, in the general blood flow.

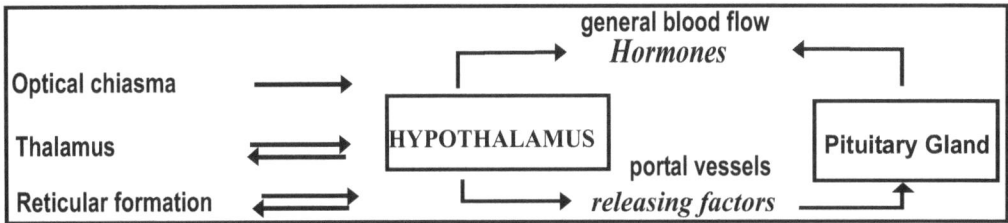

Fig. (1.7). Hypothalamus and pituitary gland.

The neuro-hormonal system use nervous cells releasing a very weak amount of transmitter to produce a large amount of hormone acting on a distant target organ thanks to hormonal amplification (Fig. **1.8**).

Fig. (1.8). The hormonal amplification.

The magnitude of CRF release is about 10^{-9} gram/h, the cortisol release 10^{-3} gram/h. Note that all the transmitting molecules are highly specific.

Pheromones are non local synapses: the pre-synaptic apparatus and the receptor do not belong to the same animal. This chemical link plays an important part in insect social behavior. Thanks to pheromones, the autonomous individual is the swarm, not the bee.

1.1.3.2. The Two Ways to Obtain Synaptic Specificity

The transmitter molecules released by a common synapse have to excite (or inhibit) some targets without acting on the other neurons. How to avoid chemical interactions between neurons in a tight bundle of parallel pathways?

1.1.3.2.1. Chemical Specificity of Heavy Transmitters

In the case of hormones and pheromones, there is one kind of transmitter molecule for each pathway (one different hormone for each step of each hormonal pathway). This chemical specificity avoids any mixing of signals. These specific molecules are often heavy peptides. In man, they are associated with the sense of smell and some other limited phenomena (control of hypotension, of stress, of pain…).

To understand the limitations of chemical specificity, we look at the sense of smell. The worm Caenorhabilitis elegans has no sense of view or hearing, just a chemical sense due to 32 sensors. Their specificity is provided by 10^3 genes for a total of 20.10^3 (5% of the whole genetic apparatus). In mouse, the number of genes governing only olfaction is 1% of the total number for the animal, a very huge number. If the human brain used only chemical specificity, it would need some 10^{11} different transmitters, a number obviously much too big for our genetic capacity. In fact, man uses less than 10^3 different heavy transmitters.

1.1.3.2.2. Structural Isolation of Synapses Using Light Transmitters

Genetic economy in complex neuronal systems is provided by the use of mainly two light transmitters (one exciting and one inhibitory) in each great area of the brain, the same two for all the neurons of the area. Then, the isolation of pathways is due to the design of the system: any transmitter molecule is destroyed or recaptured before its diffusion out of the synaptic chamber; this means definitely before the diffusion critical time. The diffusion coefficient D for light molecules in water is about 3.10^{-4} mm^2/s, the radius of the chamber is R = 10^{-3} mm, the characteristic diffusion time t = R^2/D is 3 ms. Definitely less than 3 ms is 1 ms. Thus: in light transmitter synapses, the transmitter has to be destroyed in about 1 ms.

Note: a) that 1 ms is just the duration of an usual spike. b) that auxiliary receptors located on the pre-synaptic membrane and on some cells surrounding the neuron (astrocytes) act to regulate the destruction process.

1.1.3.2.3. The Nature of Transmitters

We know today 10 light neurotransmitters (see Appendix H) and more than 100 peptidic (heavy) transmitters, the most famous of them being the opium-like peptides. Some other molecules (for instance ATP, the most common cellular fuel) sometimes reinforce the action of a transmitter.

1.1.3.3. From the Transmitter Release to the Receptor Activation

The synaptic apparatus (Fig. **1.3** before) is made of:

a)   A synaptic button ending the axon of the first neuron.

b)   A synaptic chamber: it is a little space between the first and the second neuron. Its width is about $20 \ 10^{-9}$ m, its radius 10^{-6} m.

c)   A dendritic post-synaptic membrane at the beginning of a second neuron. It contains specific receptors.

1.1.3.3.1. The Transmitter Release

The synaptic button encloses synaptic vesicles (about 100 vesicles in a neuromuscular synapse). Each vesicle is full of transmitter molecules (about 10^4 molecules/vesicle). When a spike reaches the button, the transmitter molecules are released in the chamber.

To be more precise, the pre-synaptic membrane contains voltage dependent Ca^{2+} channels. When a spike is triggered by the first neuron, it propagates along the axon till the synapse. Then, calcium ions enter into the synaptic button. Calcium acts as a messenger inducing (we do not know exactly how) the migration of the vesicles, their fusion with the post-synaptic membrane, their opening and the release of the transmitter molecules into the synaptic chamber (exocytose).

The released transmitter diffuses through the chamber and binds with receptors on the post-synaptic membrane. On the other hand, it is destroyed or recaptured (in about 1 ms).

1.1.3.4. From the Receptor Activation to the Post-Synaptic Channel Excitation

When bound to a transmitter molecule, a receptor acts directly or indirectly to open an excitatory or an inhibitory post synaptic channel.

1.1.3.4.1. Direct Coupling (Swift Synapses)

After each spike of the pre-synaptic neuron, the opening of the post-synaptic channel is an impulse (duration 1 ms). The ratio of open channels takes some constant value Ω depending only on the geometry of the synapse.

1.1.3.4.2. Coupling Through a Local Chemical Relay (Slow Synapses)

The signal on the receptor is amplified and widened by a chemical relay, which leads, directly or not, to the release of an "activator". The important fact is that the activator is slowly destroyed, with a time constant T_A of sometimes 10 s: the activator accumulates during successive spikes. Its asymptotic value is reached after a time T_A and depends on the firing rate F of the pre-synaptic neuron (Appendix I). The energy of the chemical relay is furnished by GTP, a cellular fuel similar to the more usual ATP: the receptor, when activated by a transmitter molecule, induces the splitting of a G protein, (so called to emphasize its affinity for GTP); fragments of this G proteins activate directly the post synaptic channels or activate other messenger molecules which activate the channels.

1.1.3.4.3. Deeply Acting Synapses

In the cases of circadian rhythms and of long term memorizing mechanisms (see part III), the signal of the receptor is relayed toward another synapse or to the cell nucleus; there, it induces the expression of a gene, the synthesis of mRNA, then of protein molecules which change the neuron excitability.

1.1.3.5. From the Activator Concentration to the Channel Opening

A very few number of measurements [6, 7] let think that the opening law, (bound to tend toward $\Omega=1$ for great concentrations), is the most linear sigmoid (see

Appendix J). Then, we can characterize each synapse by two parameters: a time parameter (T_A for slow synapses or 1 ms for swift synapses) and a "synaptic coupling force" Ω_{Max} defined as asymptotic value of the ratio of open channels for the maximum firing rate of the pre-synaptic neuron.

Activation of the usual receptors brings a contribution to the control potential which depends on the ratio Ω of opened post-synaptic channels. For slow synapses, Ω depends on the firing rate of the spikes of the pre-synaptic neuron; the value of for the maximum pre-synaptic firing rate characterizes the coupling force. For swift synapses, Ω does not depend on the firing rate. Weak and strong couplings lead to different functional properties.

1.1.3.6. Some General Connecting Rules

The number of outputs of a neuron ranges from 1 to 10^5. It is excited and inhibited by 1 to 10^5 upward neurons. Some empirical laws govern such complicated sets of connections, all of them suffering from many exceptions for synapses with heavy transmitters.

1)   All the synaptic buttons of a neuron release the same transmitter.

2)   All the synapses governed by a transmitter are exciting or all of them are inhibitory.

3)   A receptor can bind with only one kind of transmitter.

4)   A transmitter binds with several kinds of receptors. (For instance, acetylcholine binds with one kind of swift (nicotinic) receptors and several sorts of slow (muscarinic) ones. More: human receptors of brain, heart and bladder are slightly different molecules; thus, one can synthesize drugs acting on the bladder with reduced secondary effects on the brain and the heart.

5)   Many neurons receive signals from several exciting and inhibitory neurons, each of them with its particular transmitter.

6)   Many parts of the nervous system use preferentially a particular exciting and a particular inhibitory transmitter. For instance, the glutamate is the main exciting transmitter of the Central Nervous System CNS and GABA the inhibitory one.

1.1.3.7. Glial cells

A great number of supporting cells (the glial cells) surround neurons. Some of them (for instance the astrocytes) help to the recapture of synaptic transmitters and possibly to the elimination of unused neurons.

APPENDICES TO SECTION 1.1

A: Ionic concentrations outside and inside a neuron

B: Energetic of the Na/K pump

C: Ions, Nernst' potentials, generalized Ohm's law

D: The selectivity of channels

E: The gating laws

F: The Hodgkin Huxley axon

G: Calcium entrance during a spike

H: The more usual light transmitters

I: Activator concentration for a slow synapse

J: Opening *versus* concentration law

Appendix A: Ionic Concentrations Outside and Inside a Neuron

Table 1.1. Standard concentrations of vertebrates' neurons (unit: 10^{18} ions/cm^3).

Name	Z	Outside	Inside
Na^+	+1	86.1	13.2
K^+	+1	3.0	74.0
Ca^{2+}	+2	1.5	0
Mg^{2+}	+2	0.6	4.2
Cl^{-1}	-1	62.0	4.2
HCO_3^-	-1	16.3	50
PO_4^{-3}	-3	1.9	1.8
SO_4^{2-}	-2	0.2	3.0
Proteins	-1	8.9	30.15

Z is the ionic charge.

Appendix B: Energetic of the Na/K Pump

The main pump is a protein burning one ATP molecule (This is an irreversible exothermic reaction W_{ATP} = 11 kcal/mole or, in electric unit, 444 eV) to transport 3 ions Na^+ outside and 2 ions K^+ inside. *The work needed to transport an ion from the outside (concentration c_0, potential null) to the inside (concentration c_I, potential V) is Ze (V – V_{Nerst}) where V_{Nernst} is the Nernst's potential (appendix D). Thus, the total work is:*

$$W = -3e \ (V - V_{Na}) + 2e \ (V - V_K) = 3eV_{Na} - 2eV_K - eV$$

With the standard concentrations, the needed energy is W = 371 eV. Thus, we find a remarkably high efficiency W/W_{ATP} = 84 %.

Appendix C: Ions, Nernst' Potentials, Ohm's Law

The Problem

We look at ions in two containers, the outside one (concentration c_0, potential 0, temperature T) and the inside one (concentration c_i, potential V, same temperature T). The two containers are linked by a pipe (length L). We search the ionic current in the pipe.

The Method

The forces acting on the ions within the pipe are a) the electric field E = - Grad V and b) the collisions which tend to bring back the velocity distribution towards the thermal maxwellian distribution. Following the method of Maxwell and Boltzmann, we do not follow the motion of each ion, but we count the number F(x,v,t) dx dv of ions of a given sort standing at time t between x and x+dx and having a velocity in the range [v - v+dv].

Main Results

If μ is the ionic mobility and E the electric field within the pipe, the ionic flow Φ (ions per second and per square centimetre) obeys to the master equation:

$$\Phi/\mu + (kT/Ze)\ dc/dx - cE = 0$$

where k the is Boltzmann's constant. One has $kT/e = 25$ mV and $kT = 0.61$ kcal/mole.

Generalized Ohm's Law

If the pipe is not polarized, $E = -V/L$ and the above equation is easily integrated. We find the ionic current $\tilde{y}$ (A/cm^2) for one sort of ions:

$$\tilde{y} = - (\mu ZeV/L)\ [c_I - c_0\ \exp(-ZeV/kT)]\ /\ [1 - \exp(-ZeV/kT)]$$

For homogeneous media ($c_i = c_0$), it gives the usual Ohm's law.

Nernst' Potentials

If $\Phi = 0$ (no current), the master equation gives the equilibrium Nernst' potential:

$$V_{NERST} = -\int E.dx = (kT/Ze)\ Log\ (c_I/c_O)$$

Numerical Results

For the standard concentrations, the Table **1.2** gives the numerical values of the Ohm's currents $\tilde{y}$ (A/cm^2) *versus* the potential (mV) for the main ions.

Table 1.2. Generalized Ohm's law.

mV	-75	_50	-25	0	25	50
Na^+	-1008	-725	-477	-267	-101	+24
K^+	18	100	243	446	724	1065
Cl^-	21	58	177	346	572	857
Ca^{2+}	-40	-27	-15.5	-6.6	-2.1	-0.5

Detailed Computation

If c is the ionic concentration and T the temperature, the equilibrium maxwellian distribution is

$$F_0 = c\ f_{0x}\ f_{0y}\ f_{0z}\ \text{with}\ f_0 = \sqrt{m/2\pi kT}\ \exp(-mv^2/2kT).$$

We search here for the distribution F in an electric field E parallel to the x axis. As we look at one-dimensional flows, we can write equations depending on the only variable x. Newton's law tells us how the forces change the motion. Here, the forces are due to the electric field and to collisions between molecules. For diluted solutions, we can neglect the ion-ion collisions; we have to deal only with collisions between ions and water molecules. Then, the collision term is linear. To obtain a trend toward equilibrium, we have to write a (F-F₀) term. So, we obtain the detailed balance equation:

$$\mathbf{A} \equiv \partial\mathbf{F}/\partial t + \mathbf{v}\ \partial\mathbf{F}/\partial x + (\mathbf{ZeE/m})\ \partial\mathbf{F}/\partial v - (\mathbf{F_0} - \mathbf{F})/\tau = 0$$

where ZeE is the electric force, m the ionic mass and τ a time constant characteristic of the nature of the ion.

Definitions

For any F distribution, the concentration is $c = \int F.dv$ *(cm^{-3}), the flow* $\Phi = c <v>$ *is* $\int F$ *v.dv and the electric current density is* $\hat{y} = Ze\Phi$ *(A/cm^2) and the mean ionic energy of the x axis, which is* $\dfrac{1}{2c} \cdot \int mv^2 \cdot F \cdot dv$*, has to be equal to the mean agitation energy along the other axis, that is kT. We search for distributions F having a given c concentration, a given temperature T and which tend strongly toward 0 for infinite velocities. Note that one usually introduce the mobility* $\mu = Ze\tau/m$*.*

Useful Integrals

We look at the integrals $I_n = \int\limits_0^\infty y^n \cdot \exp\left(-ay^2\right) dy$ *. Integration by part leads to*

$I_{n+2} = \left(\dfrac{n+1}{2a}\right) \cdot I_n$ *. One finds easily* $I_1 = \dfrac{1}{2a}$ *.* I_0 *is obtained by computing* $\left(I_0\right)^2$ *in polar coordinates:* $I_0 = \dfrac{1}{2}\sqrt{\dfrac{\pi}{a}}$

The Mass Balance Equation

To obtain it, we write $\int A\, dv = 0$ and note that $\int \dfrac{\partial}{\partial t} F dv = \dfrac{\partial}{\partial t}\int F\, dv$ and that

$\int F\, dv = \int F_0 \cdot dv$ Then: $\dfrac{\partial c}{\partial t} + \dfrac{\partial \varphi}{\partial x} = 0$. In fact, after a very short transient, the flow becomes stationary. Then, dΦ/dx = 0: the flow does not depend of x.

The ionic current equation:

To obtain it, we compute the integral $\int A.vdv$. We look at a stationary problem: the $\partial/\partial t$ term disappears. Using the I_2 integral, we obtain a kT term. The $v\partial F/\partial v$ term is integrated by parts. Using the fact that $\int F_0 v \cdot dv = 0$, we see the flow Φ appear in the last term. We obtain thus the master equation c and E are functions of x:

$\Phi/\mu + (kT/Ze)\ dc/dx - cE = 0$

Replacing Φ by $\hat{y}/Ze$ and integrating this equation from x = 0, V = 0, c = c_0 to x = L, V, c = c_I, we obtain the generalized Ohm's law for any known function V(x).

Appendix D: The Selectivity of Channels

The Lateral Leakages

For each sort of ions, the current density through the membrane is

$I = D\Omega\sigma G\ \hat{y}\ (A/cm^2)$.

where $D\ (cm^{-2})$ is the density of channels, Ω the probability for a channel to be open, $\sigma\ (cm^2) = \pi R^2$ the area of an open channel and G a multiplying factor depending only of the ion properties which describes the selectivity.

The selective factor G is made of 3 terms. The first is a geometric factor: if the radius r of the ion is greater than the radius R of the channel, G = 0. Thus, Na^+ ions cannot flow through a K^+ channel. The second is due to the electric polarisation of the channel wall and depends on the ionic charge. It explains that Cl^- ions cannot flow through a K^+ channel. The third is due to the capacity of the channel wall to weakly bind to water molecules. It depends on the hydration number of the ions. The three phenomena are needed to explain that neither K^+ nor Cl^- ions flow through a Na^+ or a Ca^{2+} channel. So, K^+ channels have a very simple structure while Na^+ channels have a very complicated one.

Hints at the Mechanisms of Selectivity

The Radius R of the Central Pore of the Channel

Let R the radius of the central pore and r the radius of an ion. The effective cross-section of this channel for this ion is: $\sigma G_r = \pi(R^2 - r^2)$ if R > r and $\sigma G_r = 0$ if R <

r : *a weak diameter pore can be crossed by weak diameter ions, not by big ones. Thus, the K and the Cl channels cannot be crossed by Na, Ca or Mg. Very precise experiments using various ions (K, Cl, Na and also Li, Rb, Cs) [8] confirm this law.*

The Polarization of the Channel and the Anion/Cation Selectivity

A potential barrier inside the channel leads to an electric field different from the simple V/L. For instance, double layers near the entrance and the exit of the pore add two Dirac's functions to E. From the master equation, we find a multiplying factor for ions with charge Z_k

$$G_p = \exp (Z_k\, e\Delta V/kT)$$

With $\Delta V = 100$ mV, the ratio of anions to cations currents is multiplied by 3000. Thus, K^+ channels are not crossed by ions Cl^- and that Cl^- channels are not crossed by K^+ ions.

Weak Binding of the Channel Wall with the Water Molecules of the Hydrated Ions

Let **w** *be the binding energy of the channel wall with one water molecule and let H_k be the hydration number of K^+ ion. The binding causes a multiplying coefficient:*

$$G_h = \exp (wH_k/kT)$$

For the Na^+ channel, w = 2.4 kT. This term explains that Na^+ channels (H = 4.5) are only weakly crossed by the small K ions (H = 2.9): the filtering efficiency is exp [w (4.5-2.9)/kT] = 50. *Note that this great factor has to be associated with a small factor (due to a convenient polarization) to avoid huge space charges.*

The very simple structure of K^+ channels explains that several variants of such channels had been observed.

The Structure of the Main Channels

We sum up (Table **1.3**) the characteristics of the Na^+, K^+ and Cl^- channels explaining their selectivity. They are the radius R (10^{-10} m), the polarization ΔV (mV), the density D of channels in the squid axon membrane (cm^{-2}) and we compute also the Occupation = πR^2 D. It is the ratio of the central pores area to the total membrane area. We give the same data for the excitatory (PSe) and the inhibitory (Psi) post-synaptic channels. Note the tight distribution of the post-synaptic channels (Pse and Psi); the figures given are related to the percentage of occupation of the section of the channels by the holes and the thicknesses of the walls are not taken in account.

Table 1.3. The structure of the main channels.

	R	ΔV	D	Occupation
K^+	1.45	-110	$5.5\ 10^8$	$3.7\ 10^{-7}$
Cl^-	1.45	+110	$8\ 10^6$	$5\ 10^{-9}$
Na^+	2.32	-85 and +160	$1.4\ 10^8$	$2.4\ 10^{-7}$
Pse	2.05	0	$1.1\ 10^{13}$	$1.5\ 10^{-2}$
Psi	1.45	-30	$2.3\ 10^{13}$	$1.5\ 10^{-2}$

Appendix E: The Gating Laws

Qualitative Laws

- Chlorine channels: they are always open.

- Potassium channels: their gating loop is made of 4 identical strands. The channel is open if all the strands are open. Thus, if **n** is the probability for a strand to be open, the probability to be open for a channel is $\Omega_K = \mathbf{n}^4$.

- Sodium channels are obstructed by 3 strands of one sort and by one strand of another sort. Then, the probability for a channel to be open channels is $\Omega_{Na} = \mathbf{m}^3\mathbf{h}$ where **m** and **h** are the probabilities for the gates to be open.

For each strand, the probability to be open tends with a voltage dependent time constant $\boldsymbol{\tau}(\mathbf{V})$ toward a voltage dependent equilibrium value $\mathbf{h_{eq}}(V)$, $\mathbf{m_{eq}}(V)$, $\mathbf{n_{eq}}(V)$ with three voltage dependant time constants. For instance, $dh/dt = (\mathbf{h_{eq}} - h)/\boldsymbol{\tau_h}$.

Numerical Results

Table **1.4** hereafter gives the numerical values of the 6 functions $\mathbf{h_{eq}}$, $\mathbf{m_{eq}}$, $\mathbf{n_{eq}}$ and $\boldsymbol{\tau_h}$, $\boldsymbol{\tau_m}$, $\boldsymbol{\tau_n}$ (unit of time : millisecond).

Table 1.4. The 6 Hodgkin-Huxley gating functions.

V	heq	τh	meq	τm	neq	τn (ms)
40	0	0.67	1	0.14	0.96	1.05
20	0	0.67	0.98	0.18	0.94	1.23
0	$3\ 10^{-5}$	0.84	0.97	0.23	0.90	1.6
-20	$6\ 10^{-4}$	1.76	0.90	0.32	0.85	2.05
-40	$1.5\ 10^{-2}$	1.4	0.66	0.55	0.76	3.4
-60	0.26	5.52	0.15	0.22	0.50	5.25

Fig. (**1.9**) below shows the aspect of these 6 functions.

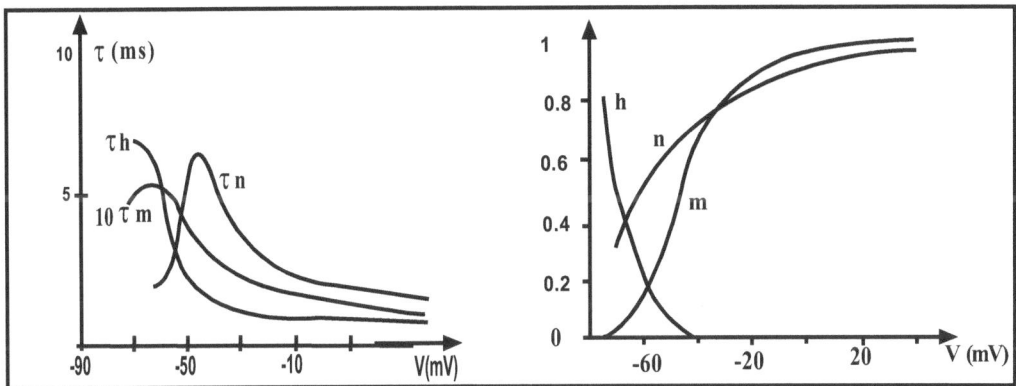

Fig. (1.9). The six Hodgkin-Huxley gating functions.

A Possible Model

We assume that each strand is a continuously deformable molecule with elastic properties (its energy depends on its length). One of its ends is fixed. The other carry an electric charge and can link with two points of the channel wall. The transitions between the two equilibrium positions are governed by the simplest kinetic law.

General Theory of Spontaneous Conformation Changes

We look at a continuously deformable molecule, its distortion being measured by some parameter x. Its potential energy is W(x). Minima of W(x) are stable configurations. We look here to a case with 2 minima with energies W_A and W_B

*for distortions. Between x_A and x_B, there is a distortion x_{Max} for which the energy W_{Max} is a maximum (Fig. **1.10**).*

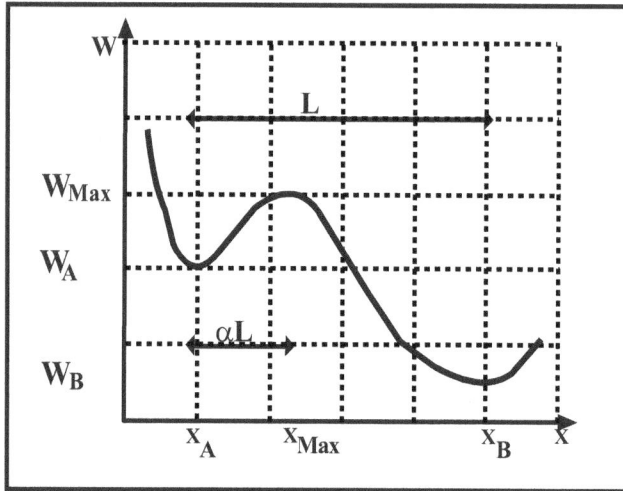

Fig. (1.10). The energy barrier governing spontaneous transitions.

From the quantum theory of thermal fluctuations, we find the rate of spontaneous transition from the stable state 1 towards the stable state 2:

Rate (A→B) = (2πkT/h) exp [(W$_A$ – W$_{Max}$)/kT] (unit : s^{-1})

where k is the Boltzmann constant, T the temperature and h the Planck's constant.

Let n_A and n_B be the number of molecules in each state ($n_A + n_B = n_{tot}$ is the invariable total number of molecules). The evolution of the population obeys to:

dn$_A$/dt = - dn$_B$/dt = n$_B$. Rate (B→A) – n$_A$. Rate (A→B)

From this, we deduce easily that the population obeys to:

dn$_A$/dt = (n$_A$eq – n$_A$) / τ

*where the equilibrium value **n$_{1Aeq}$** is given by the usual formula:*

n$_A$eq / n$_B$eq = exp [(W$_B$ – W$_A$)/kT]

*and the time constant **τ** by:*

$$1/\tau = 2\pi kT/h \exp (-W_{Max}/kT) [\exp (W_A/kT) + \exp (W_B/kT)]$$

Application to the Channel Gates

The obstruction of the central pore of a channel is due to a change of configuration of one or several proteins of the channel. The voltage dependence of the gate implies that, during its spontaneous transition, some electrically charged part of the protein travels from a weak potential area toward a strong potential one. Then the energies of the states A, B and Max are the sum of a lengthening and of an electric part. If the maximum energy place is x = αL:

$$W_A = 0 \qquad W_B = W_L + ZeV \qquad W_{Max} = W_M + aZeV_i$$

The experimental curves $m_{eq}(V)$, $h_{eq}(V)$, $\tau_h(V)$, $\tau_m(V)$ *are very accurately fitted by these formula with for instance the following values for the h gate:*

$$W_L = 10.8 \ kT, \ W_M = 13.7 \ kT, \ Z = 4 \ \alpha > 0.9$$

Note that α is almost equal to one. We would obtain such a value if the deformation energy was the sum of an elastic lengthening increasing continuously from 0 at x = 0 to W_M at x = L and of a strong short range binding of energy $W_A - W_M$ near x = L.

A Main Result: The Similitude Law

We look at two neurons, the gates of which have different strand elasticity. The equilibrium and the time constant *versus* V keep the same shape, but the equilibrium law is translated and the time constant is multiplied by a constant in such a way that slower behavior is associated to more negative potentials.

Similitude Laws

Look now at a change of the strand elasticity: W_M *becomes* $W_M + \Delta W$ *and we define* ΔV *by:* $\Delta W + Ze\Delta V = 0$. *From the above formula, one finds easily relations between the old and the new gating conditions:*

$$neq\left(V + \Delta V\right)_{new} = neq\left(V\right)_{old} \ and$$

$$\left(V + \Delta V\right)_{new} = \exp\left(\frac{Ze\Delta V}{kT}\right) \cdot \left(V\right)_{old} = \exp\left(-0.16Z \cdot \Delta V\right) \cdot \left(V\right)_{old}$$

Thus, negative ΔV leads to a slower behavior and to more negative potentials.

Appendix F: The Hodgkin-Huxley Axon

We schematise the axon by a cylinder of radius R (Fig. **1.11**). Let i be the axial density of current and I the lateral one (Fig. **A1.3**). Then, the current balance is: $\pi R^2.di = 2\pi RI.dx$, it is to say (with x in cm, R in microns):

$\partial i/\partial x = 2\ 10^4\ I\ /\ R$

The outside potential is 0, the inside potential V(x,t). Inside the axon, the usual Ohm's law applies to the axial current with the usual resistivity η of the internal medium:

$i = (10^3/\eta)\ \partial V/\partial x\ (\eta = 35.4\ \Omega.cm)$

The lateral current is due to the capacity $C = 1\ mF/cm^2$ of the membrane and to the sum of the ionic currents:

$I = C.\partial V/\partial t + I_{ions}\ (C = 1\ \mu F/cm^2)$

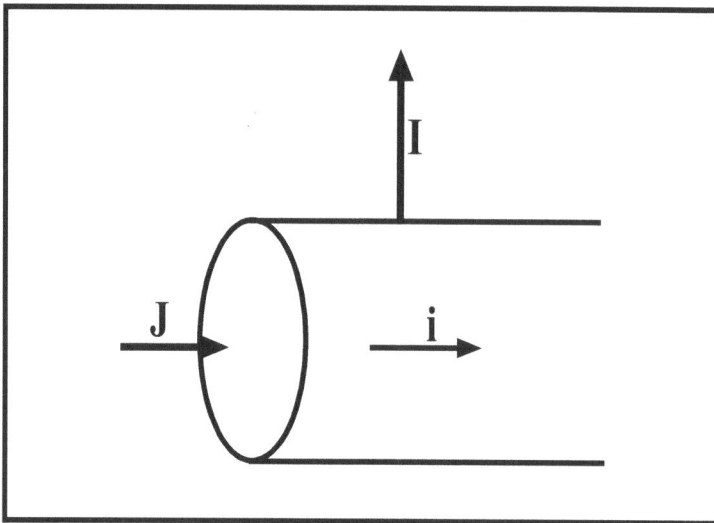

Fig. (1.11). Currents in the Hodgkin-Huxley axon.

The Ionic Currents

They are the sum of the passive transport of ions through selective channels and of the active transport of ions by the pumps. We give a special attention to Na and K passive by writing:

$$I_{ions} = I_{Na} + I_K + I_{others}$$

$$I_{Na} = g_{Na}\Gamma_{Na} (V - V_{Na}).\Omega_{Na} \quad I_K = g_K \Gamma_K(V - V_K).\Omega_K$$

$$I_{others} = g_{others} \Gamma_{oth}(V - V_{others})$$

where Ω is the proportion of open channels. For historical reasons, we introduced the correction Γ to the linear Ohm's law: $\tilde{y} = - \Gamma (\mu Ze/L) c_I (V - V_{Nernst})$ where Γ is a function of V and V_{Nernst}. The values of the constants g are:

$$g_{Na} = 120 \qquad g_K = 3 \quad g_{others} = 0.3 \qquad V_{others} = -57.7 \text{ mV}$$

Summing Up

Thus, we obtain the Hodgkin-Huxley's system:

$$(R/20\eta) \, \partial^2 V/\partial x^2 = C\partial V/\partial t + g_{Na} (V - V_{Na}) \Gamma_{Na} m^3 h + g_K (V - V_K) \Gamma_K.n^4 + g_{others} (V - V_{others}) \Gamma$$

In a rough approximation, one often takes $\Gamma = 1$.

Appendix G: Calcium Entrance During a Spike

The most important Ca channels (L channels) are found in the end part of the axons and in the post-synaptic part of muscular cells. They are gated with a very weak delay τ. Although the Nernst's potential is infinite, the generalized Ohm's law $\tilde{y}(V)$ with a zero internal concentration is well defined (but non linear) function of V (see the table in Appendix C). On the other hand, the experimental ratio Ω of open channels is well fitted by $\Omega^{-1} = 1 + \exp [-0.08 (V-0.5)]$. Then, the entering Ca current $j_{Ca} = \Omega. \tilde{y}$ is, in an arbitrary unit (Table **1.5**):

Table 1.5. Entrance of Ca^{2+} during a spike.

V (mV)	-75	-50	-25	0	25	50
j_{Ca}	0.0	0.1	0.5	1	0.5	0.1

One knows other kinds of Ca channels (T channels…) which do not play an important part.

Appendix H: The More Usual Light Transmitters

Acetylcholine: It is the neurotransmitter of neuromuscular junctions, of the ganglionic synapses of the sympathetic and parasympathetic pathways and of a great lot of places in the central nervous system. It acts on swift (nicotinic) and slow (muscarinic) receptors. Most of the opened channels are exciting ones. Myasthenia seems to be linked with acetylcholine troubles.

Glutamate: It is the main exciting transmitter of the central nervous system. It acts on swift receptors (called AMPA/kainates), on slow ones (mGluR) and on special swift receptors (NMDA).

GABA: It is the main inhibitory transmitter in the central nervous system. It acts on swift receptors ($GABA_A$) opening inhibitory Cl channels and on slow ones ($GABA_B$), which exert an inhibitory modulation. Note that some anxiolytics such as Valium® or barbiturates increase the efficiency of the $GABA_A$ channels.

Glycine: It inhibits the motoneurons of the spinal cord and the brain stem.

Dopamine: It excites the neurons of the movement controlling "substantia negra". Psychotic states could perhaps be related to excessive effects of Dopamine, Parkinson's disease to a failure of its synthesis.

Noradrenalin: The main role of noradrenalin is modulation of sleep and wakefulness. However, some noradrenergic neurons excite (α receptors) or inhibit (β receptors) targets of the sympathetic system.

Serotonin 5-HT

The main role of serotonin is modulation trough slow 5-HT1 and 5-HT2 receptors. Some anxiolytics such as Prozac® stop the serotonin recapture (increase its efficiency).

Appendix I: The Activator Concentration for a Slow Synapse

Let c be the concentration of the last messenger of the chemical chain (the activator) reaching the post synaptic channel and let γ the amount of it produced after each spike of the exciting neuron, F_{Max} its maximum firing rate and T_A the time constant for the activator disappearance. We look at a slow synapse. Then, $F.T_A$ is much greater than 1 and we can write:

$$dc/dt = \gamma.(F/F_{Max}) - c/T_A$$

The asymptotic concentration is

$$c = (\gamma.T_A).(F/F_{Max})$$

From now, we choose units such that c = 1 if Ω = 0.5. Then, $(F_{0.5}/F_{Max}) = 1/\gamma.T_A$

Appendix J: The Opening *versus* Concentration Law

A simple chemical equilibrium:

H activator moles + closed channel $\leftrightarrow$ open channel would lead to the law:

$$\Omega = c^H / (1 + c^H)$$

In the case of co-operative effects, Hill's number H has not to be an integer. Experiments on acetylcholine and GABA gave $1.5 < H < 2$.

A Functional Optimisation of the Linearity of Ω as a Function of c

Searching for a law Ω *vs* c almost linear for weak c values, we find **H = 1.7**:

1.2. Physics and Chemistry of Any Kind of Neuron

1.2.1. Homogeneous Cylindrical Parts of Every Possible Neuron

Main results: The sensitivity of a neuron depends on the density of channels in the cell body, the scale of its firing rate of the ionic concentrations.

Here begins the original part of our work: classifying the different kinds of neurons. Obviously, Hodgkin-Huxley's ionic mechanism governs all of them. But all neurons (and all the parts of a neuron) are not alike. Channel densities are not the same in the cell body (the most sensitive part of a neuron) and in the axon (just a transmitting line). They have not to be the same in different neurons. On the other hand, ionic concentrations and Nernst potentials, which have to be homogeneous inside a neuron, have not to be the same in different neurons. This Section is devoted to the search of the authorized differences.

1.2.1.1. Variations of the Channel Densities

A change in the membrane channels density would change the neuron sensitivity (a neuron with a very weak Na^+ channels density would not be excitable). We conceive two independent kinds of such changes (Appendix A):

a) Either to multiply the densities of Na^+/K^+ pumps and of Na^+, K^+ and Cl^- channels by the same factor. Then, neither the Nernst potentials nor the rest potential are changed. This parameter p_1 governs the neuron sensitivity.

b) Or to change only the K^+ (multiplicative parameter p_2) and Cl^- densities in such a way that the rest potential does not change. (p_1 and $p_2 = 1$ for the standard neuron).

1.2.1.2. Variations of the Ionic Concentrations

1.2.1.2.1. Adjustment of the Gates Stiffness

Some neurons (for instance those of the optic nerve) reach a firing rate of $1500\ s^{-1}$ (an interval between spikes of 0.7 ms) Such neurons cannot obey to the standard kinetics of Na^+ channels for which τ_h and τ_n lead to a spike width much greater than 0.7 ms. We have to look at channel with a faster gating.

The Na^+ channel is such a complicated set of proteins that there is only one sort of it. But, after looking at the physics of gating (Appendix E of Section 1.1), we assume that changes in the channel stiffness translate the curves h_{eq} and m_{eq} *versus* V by a factor ΔV and multiplies their time constant τ by a factor Θ such as $\Theta = \exp(-0.16\Delta V)$.

This relation is supported by various measurements of τ_m *versus* V for very different excitable cells (squid axon, frog motor fibre, frog sensitive fibre and mammalian heart Purkinje's cell), after being corrected with appropriate Θ and ΔV, superimpose perfectly.

1.2.1.2.2. Adjustment of the Ionic Concentrations

Efficient working of the cell implies that the rest potential must be at the foot of the m_{eq} *vs* V curve so that the Na^+ leakage at rest is very weak (see figure in

Appendix E of Section 1.1). To fulfil this condition when the channel stiffness is changed, the inside ionic concentrations **c** (and the densities of auxiliary pumps) must also be changed in such a way that the rest potential is translated of ΔV. But the concentrations have to respect several other constraints (Appendix B): inside and outside electro-neutrality, equilibrium of inside and outside osmotic pressures, Na^+/K^+ pumps working near their maximum power, rest potential greater than the K^+ Nernst potential (to allow inhibitions).

1.2.1.2.3. Application to High Frequency Neurons

All these conditions are fulfilled by the standard concentrations. Slight concentrations changes lead to swifter neurons. They require increasing the density of active pumping. Stronger changes would require unrealistic pump density.

Characteristics of high frequency neurons (inside concentrations, Nernst potentials, rest potential, Θ factor) are summed up in Table **1.6**. Experimentally, the range of the rest potential of most of the neurons is [-70, -60 mV].

Table 1.6. Ionic properties of high frequency neurons.

cNa^+	cK^+	V_{Na}	V_K	V_R	Θ
13.2	74.0	46.9	80.1	70.0	1.0
12.5	74.7	48.2	80.4	65.7	0.5
12.2	75.1	48.9	80.6	62.2	0.33

1.2.1.2.4. Application to Heart Pace-Makers (The Purkinje's Cells); Interest of a Blood Barrier

They are excitable cells. Their main property is to fire spontaneously with a very low firing rate (about 1 s^{-1}). To obtain a firing rate range from 40 to 120 spikes per minute, we expect a widening of time $\Theta = 210$ which leads to $\Delta V = -33$ mV and to a rest potential $V_R = -103$ mV.

How to obtain these numbers? The heart excitable cells are not protected by a blood barrier. Thus, the external medium has a high concentration of neutral molecules and the osmotic pressure is high. It is equilibrated by a very high inside potassium concentration $c_K = 440$. Other inside concentrations are $c_{Na} = 78$ and $c_{Cl} = 2$.

The Nernst potentials in this case are $+15$ mV for Na^+, -111 mV for K^+ and -98 mV for Cl^-. The rest potential is about -100 mV which is the expected order of magnitude.

So, we discover an important advantage of the blood-brain barrier: by filtering most of the heavy neutral molecules, it decreases the osmotic pressure of the outside medium of neurons, inducing an increase of the spontaneous firing rate by a factor greater than 200.

1.2.2. Any Part of Any Neuron

Main results: The general relations between inputs and outputs of any neuron.

1.2.2.1. Post-Synaptic Channels and Post-Synaptic Potentials

We know excitatory and inhibitory synapses. Their opening Ω depends (directly or not, see Section 1.3) on the transmitter release by the previous neuron. The maximum value of Ω for each synapse characterizes the synaptic coupling.

The properties of the two types of channels have been measured [9].

1.2.2.1.1. Excitatory Post-Synaptic Channels

When open, they allow passive transport of K^+, Na^+ and Cl^-, but not of larger ions (see Appendix 5). The zero current is obtained if:

$$V_{Exc} - V_R = U_{Exc} \qquad \text{with } U_{Exc} = +32 \text{ mV}$$

For a fully open common synapse (with a standard number of 3.10^5 channels), the synaptic current is:

$$J_{Exc} = 3.7 \ 10^{-3}(V-V_{Exc}) \ \mu A/\text{synapse}$$

1.2.2.1.2. Inhibitory Post-Synaptic Channels

When open, they allow a strong passive transport of K^+ ions, a weak passive transport of Cl^- ions, but do not allow the transport of Na^+ ions. The zero current is obtained if:

$$V_{Inh} - V_R = U_{Inh} \qquad \text{with } U_{Inh} = -10 \text{ mV}$$

For a fully open standard synapse (7.10^5 channels), the synaptic current is:

$$J_{Inh} = 1.2 \ 10^{-3}(V-V_{Inh}) \ \mu A/synapse$$

1.2.2.2. The Control Potential

The effects of all the exciting and inhibiting synapses of a neuron add in a complicated way to create a control potential V_C which depends on the whole neuron properties. It is convenient to compare V_C to the rest potential V_R by writing

$$V_C - V_R = U_C$$

Appendix C gives the addition law. Here, we only note:

a) That the range of U_C is [-10, +32 mV] (The extreme values are never reached).

b) That neurons with a big radius need a greater number of synapses than neurons with a little radius.

c) That one excitatory synapse is more effective than an inhibitory one.

1.2.2.3. The Triggering Area

The HH axon is a homogeneous cylinder. The entrance of real neurons (Fig. **1.1**) is most of the time a dendritic tree, which excites a cell body, an axon and a terminal arborescence. The radius of the cell body is greater than the radius of the axon. The dendritic tree creates the control potential (Uc less than 32 mV). We observe that, except in neurons with very long cell body (pyramidal cells), the spikes (U_{spikes} much greater than 32 mV) are always created at the same place, the triggering area. This place has obviously to be more excitable that the others. It will be so if it is made with the same material as the other places (the dendritic tree, the axon) but owns a greater density of channels. (In pyramidal cells, there are several competitive triggering areas).

1.2.2.3.1. Spikes Do Not Perturb the Control Potential

The connections between the dendrites and the cell body has a discontinuous profile, the impedance is badly adapted. That remark means that a signal cannot go from the cell body backward the dendrites. The strong damping of backward propagation of spikes in the dendritic tree was experimentally observed [10] in the cerebellum Purkinje's cells. Thus, U_C depends only on the opening of the post-synaptic channels and governs the self-oscillating behavior of the cell body potential U. (The relation between U_C, U_{Spike} and U is discussed in Section 1.3)

1.2.2.4. Propagation of a Spike Along an Axon

The propagation of little potential oscillations along the axon is strongly damped. On the contrary, spikes propagate without shape deformation. The propagation velocity v does not depend on any factor except the neuron nature. Unmyelinated fibres behave like Hodgkin Huxley cylinders. The computed velocity (Appendix D) is $v = 1.7\sqrt{R}$ (for standard concentrations) when the experimental law [11] is **$v = 2.4\sqrt{R}$** (units: v in m/s, radius R of the axon in microns, range [1 - 10 m/s].

The contrast between strongly and poorly excitable parts of a neuron is maximized in long myelinated fibres. The isolating coating prevents ionic escape along the axon. So pumping (and energy expense) is very weak. As a passive propagation tends to decrease by expanding the spike shape, periodically spaced active areas (the Ranvier's nodes) act as sub-marine cable repeaters. For myelinated fibre, the theory and the experiment lead to a velocity proportional to the radius (range [3 - 100 m/s]). Some neuronal sub-systems (for instance the pain circuitry) use the difference between fast and slow transmission of spikes.

1.2.2.4.1. There is No Backward Signal

The profile of transition from the cell body to the axon is smooth. Moreover, the delays in gates opening forbid a reflection of the spike at the end of the axon. Thus, the axon acts as very well matched impedance.

1.2.2.5. Transient Neurons

When submitted to a constant excitation, the firing rate of many neurons is constant with time while the firing rate of many others decreases (Fig. **1.12**). We call the last ones <u>transient neurons</u>. For instance, many cells of the visual systems of men and frogs are transient; then, the eye is more sensitive to moving targets than to static ones.

This spike frequency adaptation (Appendix E) is due to the working of Ca^{2+} dependent K^+ channels located in the cell body and the axon of transient neurons (and different from the usual voltage dependent K^+ channels). Ca^{2+} ions appear inside the cell during each spike; they induce the opening of these special K^+ channels [12] while their spontaneous closure is slow. The number of open channels is a cumulative phenomenon [13] increasing with the spikes firing rate. The effect of the K^+ leakages is described by adding a negative term to the control potential.

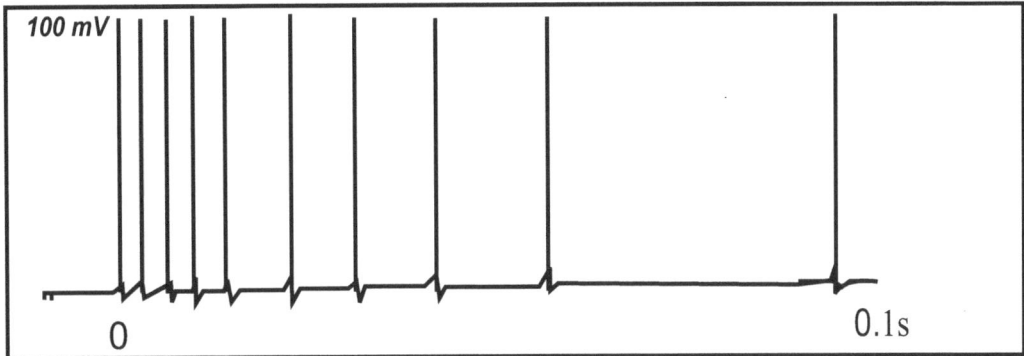

Fig. (1.12). Firing of a transient neuron.

1.2.2.6. Circadian Neurons

In man, circadian rhythms are governed by 10^4 specialized neurons located in the suprachiasmatic nuclei, a part of the hypothalamus. The control of the cell excitability by the genetic apparatus is spontaneously oscillating with a period slightly greater than one day. All neurons own two circadian genes, almost the same in man and drosophila fly, which induce the synthesis of two circadian proteins on the ribosomes of the cytoplasm. They form dimers which are stored. When their concentration is high enough, some of the dimers dissociate and are transported in the nucleus where they bind with the gene promoter, stopping its transcription. In normal neurons, the circadian proteins yield is weak and the feedback loop leads to a stable steady state. In circadian neurons, the yield is strong and the over-regulated device has a spontaneously oscillating behavior: when the protein concentration is high, the gene expression is stopped for a time equal to the protein-promoter bound state lifetime; when the protein concentration is low, its synthesis occurs during a time due to the storage of the proteins as dimers [14].

APPENDICES TO SECTION 1.2

A: Non standard densities of channels

B: Non standard ionic concentrations

C: The addition of post-synaptic potentials

D: The propagation of a spike along an axon

E: Transient neurons

Appendix A: Non Standard Densities of Channels

The standard channels populations are described by the 3 coefficients g_{Na}, g_K and g_{other}. Can we arbitrary change these 3 numbers without changing the rest potential?

a) We can (within some limits) multiply the three numbers by the same factor p_1:

$$g_{Na} \rightarrow g'_{Na} = p_1 g_{Na} \qquad g_K \rightarrow g'_K = p_1 g_K \qquad g_{other} \rightarrow g'_{other} = p_1 g_{other}$$

*b) Without changing the density of Na channels, we try now to change the density of K channels without changing the rest potential. For the standard axon at rest (V = -70 mV), the total current is zero (Goldman's law). The HH equations give then **neq** = 0.312, **meq** = 0.042, **heq** = 0.608. From these numbers, we find the partial ionic currents at rest:*

$$I_{Na} = -0.6317 \ \mu A \ / \ cm^2 \ I_K = 3.4553 \ I_{other} = -2.8236$$

As expected, their sum is zero. To keep this zero condition when changing the density of K channels, we have to change also the density of Cl channels to keep constant the sum $I_K + I_{other}$

$$g_{Na} \rightarrow g'_{Na} = g_{Na} \qquad g_K \rightarrow g'_K = p_2 g_K \qquad g_{other} \rightarrow g'_{other} = (1.224 p_2 - 0.224) g_{other}$$

Thus, the possible channels densities are entirely described by the two coefficients p_1 and p_2.

Appendix B: Non Standard Ionic Concentrations

They have to respect two absolute physical laws:

Inside and outside electro-neutrality

Σ **(Zc)** inside (sum of inside electric charges) = 0 and Σ **(Zc)** outside = 0

Equilibrium of inside and outside osmotic pressures:

Σ **c** inside = Σ **c** outside

(sum of inside concentrations = sum of outside concentrations)

Moreover, they have to respect two optimizations caused by the evolution:

they must keep the Na^+/K^+ *pumps working near their maximum power* (Appendix B of Section 1.1):

$3V_{Na} - 2V_K - V_R = 371$ **mV**

They must *keep the possibility of both excitations and inhibitions*, that is a rest potential greater than the potassium Nernst potential:

$V_R \geq V_K + 10$ **mV**

Appendix C: The Addition of Post-Synaptic Potentials

We look at a Hodgkin-Huxley cylinder beginning at x = 0 (Fig. **1.11**). The entering current J is the sum of the post-synaptic currents. So, it depends on the numbers $N_e\Omega_e$ and $N_i\Omega_i$ of exciting and inhibitory open and functional synapses and on the local potential V_C (or, more conveniently, the distance $U_C = V_C - V_R$ between the actual and the rest potentials).

Let R be the radius of the cylinder, f and ω the damping and the pulsation of little oscillations of the HH potential (see Appendix A to Section 1.3). We obtain after some calculations (R in micron, f and ω in ms^{-1}):

$$U_C = 32\frac{N_e\Omega_e - 0.1N_i\Omega_i}{N_e\Omega_e + 0.324N_i\Omega_i + K}$$

$$with \quad K = 0.5R^{3/2}(f^2 + \omega^2)^{1/2}$$

Note the balance between the excitations and the inhibitions and the very great dependency of K on the neuron cell body radius R.

Hints at the Calculation

From the potential/current relation for post-synaptic channels, we can write:

$$J = \alpha\,(U_C - 32) + \beta\,(U_C + 10)$$

where α and β are two numbers.

We schematise the neuron as a homogeneous cylinder with the radius R of the cell body (unit: micron) and we search for a stationary potential $U(x) = U_C \exp(-x/\lambda)$ inside the cylinder. Let $i(x)$ be the density of axial current, $I(x)$ the radial density of current. From the HH equations, we obtain $I = \mu\,dU/dx$ and $dI/dx = i$ where the lateral leakage i is computed in stationary conditions ($m = m_{Eq}$) and at the first order in U. Then, $i = vU$ where μ and v are two numbers. The solution of these equations is

$$\lambda^2 = \frac{\mu}{v} \quad I = \mu U_C \exp(-x/\lambda)\; and\; J = I(\,x = 0)$$

The numerical value is $\lambda = 380\,[R\,/(\,f^2 + \omega^2)]^{1/2}$ (f and ω in ms^{-1}; λ and R in microns), an always great value: the control potential is not damped by the distance.

Appendix D: The Propagation of a Spike Along the Axon

For fibres without myelin, we start from the HH equations:

$$(R/20\eta)\,\partial^2 U/\partial x^2 = C\partial U/\partial t + g_K\,n^4\,(U - U_K) + \cdots$$

And we search for an self similar solution (that is a solution $U(x,t)$ depending only of the variable x-vt. To simplify, we look only at the top of the spike. It is a top, hence $\frac{\partial U}{\partial t}=0$. The potential U at the top is the Nernst potential of Na, hence $I_{Na} = 0$. Assuming that the potassium gates had not the time to open (n has its rest value) and $n^4 = 0.0095$. If θ is the spike duration, $U \approx U_{Na}\,[1 - (2t/\theta)^2]$ and $\partial^2 U/\partial t^2 = 8\,U_{Na}\,/\,\theta^2$. We want to evaluate v in the expression $v^2\,\partial^2 U/\partial x^2 = \partial^2 U/\partial t^2$. With the above values, we find:

$$v = 1.7\,(R/\partial_1\,\partial_2\,\Theta^2)^{1/2} \quad (units:\; v\; in\; m/s,\; R\; in\; microns).$$

For myelinated fibres, *the axon is made of segments of length l isolated by the myelin; the propagation within them is a simple diffusion. They are separated by Ranvier's nodes, which own channels and restores the spike shape.*

Appendix E: Transient Neurons

In transient neurons, Ca^{2+} ions increase the ratio of open transient K^+ channels; between spikes, this ratio decreases slowly. Let q be, in a well-chosen unit, the number of open special **K^+** channels. Then:

$T_D \, dq/dt = F/F_D - q$

where T_D is the time constant for the decrease of effective control potential (often 0.1 s) and where F_D is a constant that governs the value of the neuron firing rate after decrease.

With K^+ leakages proportional to q, the effective control potential **U_{Eff}** (which governs the master equation: the firing rate F is a function of U_{Eff}) differs from the usual control potential **U_C** (which depends only on the synaptic activity) proportionally to the number of open channels.

$U_{Eff} = U_C - q$

Application

Let $F = \psi(U)$ the firing rate versus the effective control potential and $U = \Phi(F)$ the inverse function.. At time t=0, q=0, F=F_0. Hence: **$\Phi(F_0) = U_{Eff}(t=0) = U_C(t=0)$**

After a time of the order of T_D, dq/dt tends toward 0 and $q = F/F_D$. Hence, for t infinite:

$\Phi(F_\infty) = U_{Eff}(t=\infty) = U_C(t=\infty) - F_\infty/F_D$

For damped neurons, these equations have always a solution and this solution is near of:

$U_{Eff} = U_{Thr} + F_\infty/F_D$

For oscillating neurons, the function Φ is defined only if $F > F_{Min}$ such that $U_{OFF} = \Phi(F_{Min})$. For $F_\infty < F_{Min}$, the above formulas cannot be used. We observe then *relaxation oscillations: q increases, F decreases till F_{Min} and stops, q increases slowly with a time constant T_D. New spikes occur when $U_{Eff} = U_C - q$ reaches the threshold value U_{ON}.*

1.3. A Kit of Simple Input-Output Relations

1.3.1. The Nervous Message

Main result: The firing rate is the nervous message.

In the chip of a computer, electronic phenomena support the processing of purely abstract (digital) messages. In the same way, the chemical and electric phenomena of the various neurons support the processing of purely abstract nervous messages. The function of nervous processing is to link sensorial inputs to memory, thought and muscular outputs. The global process results from the individual transformations of messages going through a neuron: the functional working of a neuron is described by its input-output relations. (For instance, the output is the quantitative sum of two inputs). Thus, we have: a) to define precisely what is a nervous message; b) to find how the input-output relation of a neuron is linked to its electric and chemical properties; c) to catalogue the most useful input-output relations; and d) to present a kit of simple neuronal sets carrying out simple logical or quantitative operations. *The nervous message is the average firing rate F of spikes.*

The nature of the nervous message was discussed during many years. Looking at the analogies with an industrial transmission line finally solved this problem. Transmitter release occurs only when a spike reach the end of the axon. The shape of the spike does not depend on any circumstantial phenomenon. Thus, little perturbations of the control potential are not transmitted; the only phenomenon measured at the end of the axon is the occurrence of a spike.

There are only two possibilities: either the required information is contained within the relative timing between spikes of the same sequence or, in the simplest way, this information is in the number of spikes transmitted during a same amount of time. The first option is not acceptable: measuring time and decoding the message would imply that some kind of clock is available in both neurons; there is not such a clock. If the studied neuron and the following one are linked by swift synapses, the opening of the post-synaptic channels of the following neuron is very short 1 ms). Then, the useful signal is the occurrence of a spike. If the studied neuron and the following one are linked by slow synapses, the useful nervous signal is the average firing rate of spikes F, averaged over the synaptic delay. Note that **F** is an increasing function of the control potential Uc.

Use of averaged firing rate is a coding method different from frequency modulation (which use continuous signals and not impulsive ones) and from the digital coding used in a computer (which use yes-or-no messages at times fixed by a clock).

To summarize: The *nervous message is the average firing rate F of spikes or the occurrence of an isolated spike.*

1.3.2. A Functional Description of Neurons

Main result: Linear or all-or-none behavior of synaptic couplings.

1.3.2.1. Specific Structural Parameters of a Neuron

This paragraph is a short reminder of section 2. The specificity of any neuron is characterized by a little number of structural parameters.

The *triggering area* (the cell body) depends on 4 parameters: its radius **R**, the parameters **p₁** and **p₂** governing the densities of its channels and the multiplicative time factor **Θ** associated with non standard ionic concentrations.

As spikes propagates along the axon with a velocity depending only on the axon radius, the *propagation* is described by 1 parameter, the delay τ_{Axon} (range from 0.01 to 1 s for nerves going to or from the legs, delay neglectible for neurons of the central nervous system).

Each *synapse* between two neurons is characterized by 4 parameters:

- The number n_e (or n_i) of exciting (or inhibiting) of functional post-synaptic receptors (which we compare often to the number of receptors of a standard synapses: $N_e = \dfrac{n_e}{n_{e0}}$).

- The time life **T$_A$** of the activator (T$_A$= 1 ms for swift neurons).

- The amount **γ** of activator produced by a spike.

(Remind that a neuron can excite and be excited by several others; each of these connections is described by its set of numbers N, T, γ).

Transient neurons are characterized by 2 parameters: the time constant T_D for the decrease of effective control potential and by a frequency F_D that governs the final value of the neuron firing rate.

1.3.2.2. A Qualitative Description of the Functional Phenomena

(Remind that we use henceforth the distance **U** to the rest potential $U = V - V_R$)

From the solutions of the HH equations, we obtain the description below. The control potential U_C depends on the synaptic excitations, which varies with time. During a first phase, the Na^+ channels of the cell body (the triggering area) are almost closed. The potential U of the triggering area is not equal to U_C but is driven by it. When the neuron is not firing, U follows U_C with little oscillations (Fig. **1.13**). If U reaches some threshold U_{Thr}, Na^+ channels suddenly open. This is the beginning of a self-governed (an explosive) process, the spike, which propagates along the axon. It evolves in an imperturbable way. Its duration is $\theta_{Spike} = \Theta$ ms (1 ms for neurons wih standard concentrations). The end-of-spike conditions are $U = U_K$ (Nerst potential for K^+) and $dU/dt = 0$.

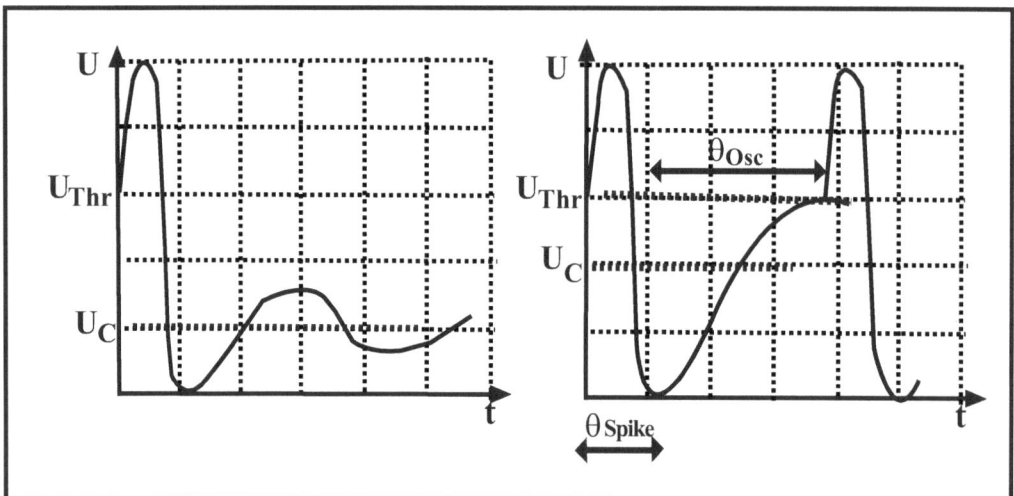

Fig. (1.13). Effect of a constant control potential after a first spike: (left: weak U_C; right: strong U_C).

1.3.2.3. Specific Functional Parameters of a Neuron

1.3.2.3.1. From the Control Potential Uc to the Firing Rate F

From the control potential to the potential U of the triggering area:

From the Hodgkin-Huxley equations, one finds the damping coefficient **f** and the pulsation ω for the weak oscillations of the potential U during the first phase. These parameters are specific functional parameters. They are computed from the specific structural parameters (Appendix A). Note that, in some neurons, U tends toward U_C in a doubly damped way (with two damping parameters f_1 and f_2). Henceforth, we will call "fully damped neurons" the neurons with f_1 and f_2 and oscillating neurons the neurons with f, ω.

1.3.2.3.2. The Threshold Potential

The parameter U_{Thr} is a specific structural parameter which governs the triggering of a spike. It is computed from the specific structural parameters (Appendix A).

1.3.2.3.3. Firing Rate Versus Control Potential

For a constant control potential, the firing rate depends only of U_C, f, ω and U_{Thr} (Appendix B). Fig. (**3.1**) shows how to compute it: a first spike ends (at time θ_{Sp}). The end-of-spike conditions are initial conditions for the following oscillating phase, (which has an overshoot for oscillating neurons). U reaches the threshold value (beginning of the following spike) at time $\theta_{Sp} + \theta_{Osc.}$. The firing rate is:

$F = 1000/ (\theta_{Sp} + \theta_{Osc})$ (F expressed in Hz and θ in ms)

The response of a transient neuron can be computed in the same way.

Starting and stopping conditions: Characteristic potentials U_{OFF} and U_{ON}

We look now to initial conditions of the oscillating phase different from end-of-spike ones.

If U_C increases slowly, the overshoot is negligible, $U = U_C$ and firing begins when $U_C = U_{Thr}$

OFF potential: we look at a firing neuron. U_{OFF} is the constant excitation such that the potential U(t) just does not reach the threshold value.

A firing neuron stops if $U_C < U_{OFF}$

ON potential: we look now at a neuron at rest submitted to a step-like excitation. The overshoot triggers a first spike if $U_C = \mathbf{U_{ON}}$

For fully damped neurons, $U_{Thr} = U_{ON} = U_{OFF}$; for oscillating neurons:

$U_{Thr} > U_{ON} > U_{OFF}$ (Appendix C).

1.3.2.3.4. Critical Impulse Height H_{ON}

We look at the value $\mathbf{H_{ON}}$ of an impulse of the control potential 1 ms wide (or Θ ms wide for non standard concentrations) such that it triggers a spike in a neuron at rest (Appendix D).

From the firing rate of a neuron to the control potential of the following neuron

1.3.2.3.5. Linear or Yes-or-No Coupling

A slow synapse is excited by a first neuron, maximum firing rate F_{Max}. The opening of the post synaptic channels of the following neuron (Appendices E and F) depends on the activator lifetime (range 20 ms to 10 s) and on the number γ of activator molecules released after each incoming spike.

Fig. (**1.14**) shows that the product γ^*T_A is a <u>linearity index</u>: if $\gamma^*T_A \gg 1$, the synapse behaves as a yes-or-no system with $\Omega > 0.5$ for $F > 0.1\ F_{Max}$. If γ^*T_A is close to 1, the opening Ω is proportional to the exciting firing rate with a maximum Ω_{Max} about 0.5. If $\gamma^*T_A \ll 1$, the channels never open.

1.3.2.3.6. Strength of Synaptic Connections

The control potential U_C generated by N exciting synapses, all of them with the same opening Ω, depends on Ω, N, f, ω and on the radius R of the triggering area. We choose as <u>synaptic strength index</u> the number N_1 of standard synapses needed to obtain the threshold $U_C = U_{Thr}$ with fully opened synapses ($\Omega = 1$) and standard radius $R = 3\mu$. (Appendix G). For other radius, N_1 is proportional to $R^{3/2}$.

Let N be the actual number of synapses. If $N < N_1$, the coupling is too weak, the synapses cannot trigger the firing of the neuron. For yes-or-no coupling, the maximum value of Ω is about 1: the connection is usable if N is slightly greater than N_1. For linear coupling (Fig. **1.14**), the maximum Ω value is about 0.5 and gradual response implies a large usable range, for instance from $\Omega = 0.05$ to $\Omega = 0.5$: an efficient linear coupling is such that $\Omega_{thr}*N > N_1$ (Number of usable channels multiplied by the opening ratio greater than N_1), that is $N > (1/0.05)\ N_1$ or $N > 20\ N_1$. These conditions lead to use neurons with a weak threshold and to a sufficient number N of synapses (which has to increase with the cell body radius R).

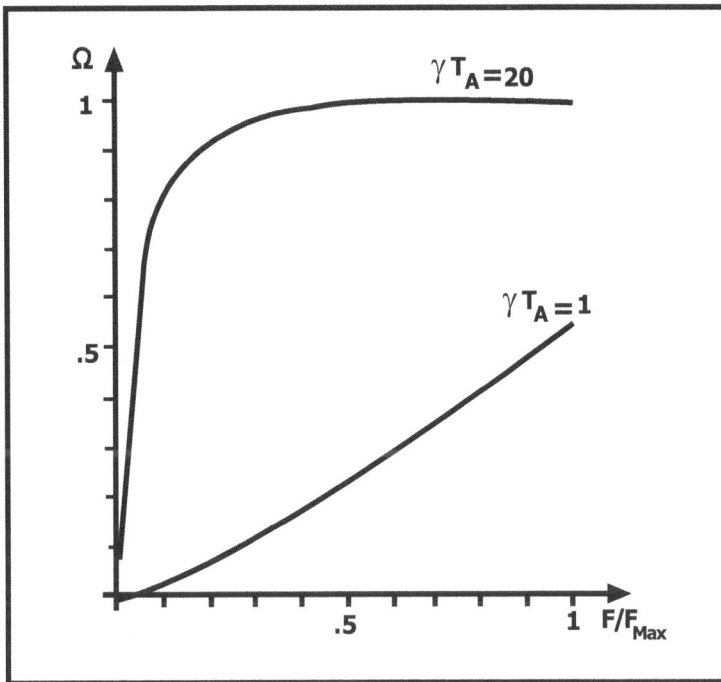

Fig. (1.14). Linear and yes-or-no responses.

1.3.3. Typical Neurons

Main results: Short or long duration, linear or all-or-none typical responses.

<u>1.3.3.1. Three Oscillating Neurons</u>

<u>1.3.3.1.1. Characteristic Parameters and Firing Rate Versus Control Potential Relation</u>

We call them aperiodic, periodic and spontaneous neurons (see below). They are defined by the structural and functional parameters of the Table **1.7** and by standard concentrations ($\Theta=1$). The voltage unit is mV, the frequency unit ms^{-1}.

Table 1.7. Functional parameters of typical neurons.

	p_1	p_2	ω	f	U_{Thr}	U_{ON}	U_{OFF}	H_{ON}
Aperiodic	1	1	0.50	0.21	6	4.7	2.6	18.1
Periodic	2.4	0.75	0.67	0.15	3.2	2.1	-1.2	6.3
Spontaneous	3	0.5	0.54	-0.10		2.4	-5.6	

Note 1: The negative value of f for the spontaneous neuron: any little perturbation is sufficient to trigger its activity if it is not strongly inhibited. Then, U_{ON} and H_{ON} are not defined.
Note 2: The negative value of U_{OFF} for the periodic neuron. Once firing, it continues to fire without excitation.

Fig. (**1.15**) shows the firing rate *versus* the control potential (Appendix H).

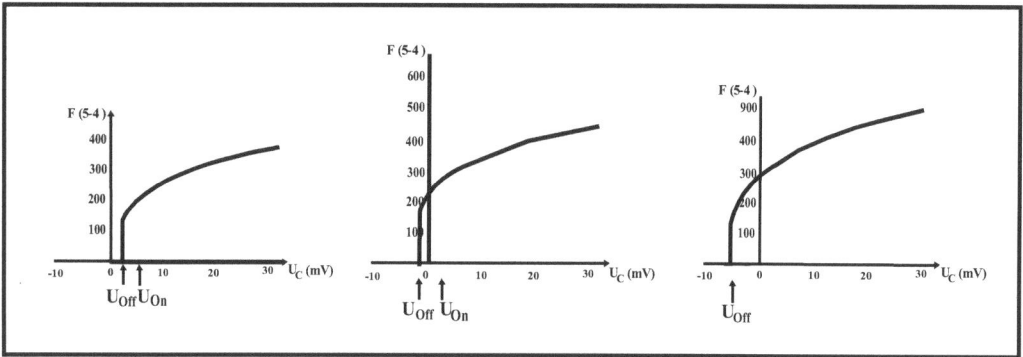

Fig. (1.15). Firing rate of typical neurons.

1.3.3.1.2. Start and Stop Behavior

Aperiodic Neuron: Without external excitation, it stays at rest: U = 0. When submitted to a sufficient excitation, the neuron fires with firing rates in the range [140 - 350 s^{-1}]. When the excitation stops, its activity stops (Fig. **1.16**).

Periodic Neuron: It is similar to an ON/OFF system (Fig. **1.17**).

When unexcited, it stays at rest. When submitted to a sufficient excitation, it starts firing with firing rates in the range [180 - 440 s^{-1}]. When the excitation stops, it goes on firing with a slightly weaker firing rate till an inhibition stops it.

Spontaneous Neuron (Pace Maker): It is always firing with firing rates in the range [150 - 420 s^{-1}] except when it is strongly inhibited (Fig. **1.18**).

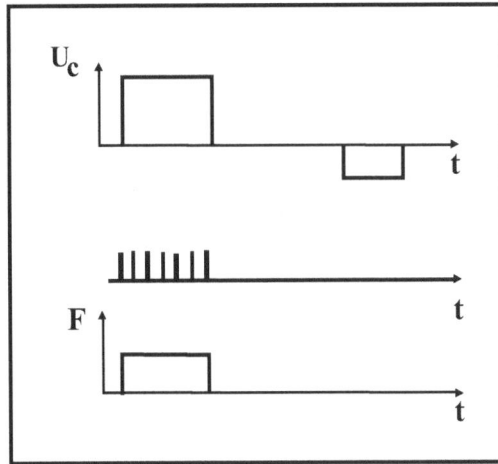

Fig. (1.16). Response of an aperiodic neuron to an excitation and then to an inhibition. Top: control potential U_C *versus* time. Middle: aspect of the spikes as they would be seen on a screen. Bottom: firing rate of spikes *versus* time.

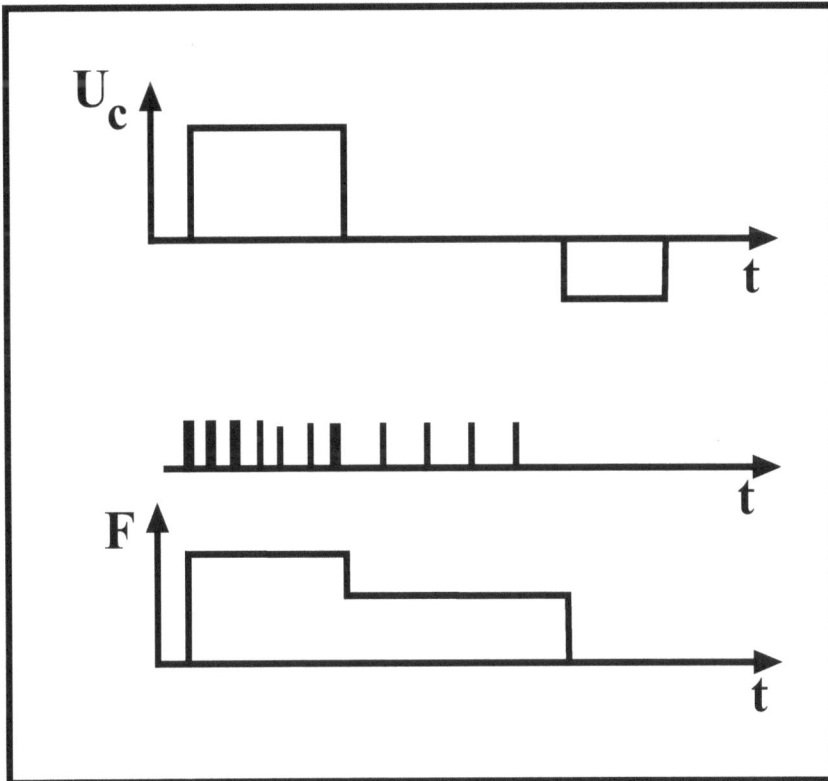

Fig. (1.17). Response of a periodic neuron to an excitation followed by an inhibition.

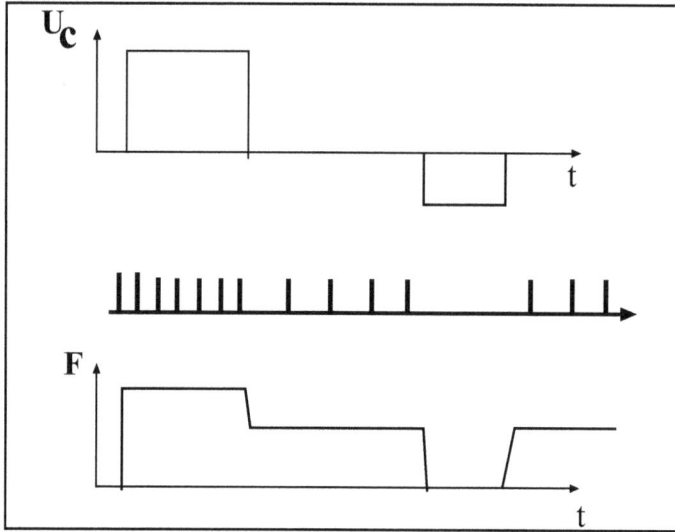

Fig. (1.18). Response of a spontaneous neuron to an excitation followed by an inhibition.

Transient Neuron: If submitted to a constant excitation and if $F_{\infty} \rangle F(U_{OFF})$, the firing rate decreases from F_0 to F_{∞} (Fig. **1.19**).

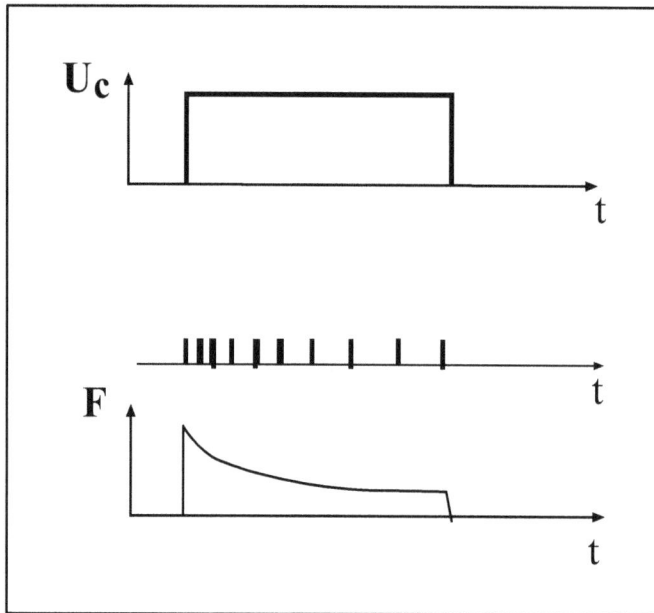

Fig. (1.19). Response of a transient neuron to a constant excitation.

If $F_\infty \langle F(U_{OFF})$, relaxation oscillations are observed (see appendix E of the previous Section); the neuron fires while U_C decreases from U_{ON} to U_{OFF}, then is silent while U_C increases from U_{OFF} to U_{ON}.

1.3.3.2. Two Typical Fully Damped Neurons

1.3.3.2.1. Characteristic Parameters and Firing Rate Versus Control Potential Relation

We look at a swift neuron fitting well the optic nerve and to a slow neuron fitting well the motoneurons. Both of them have slightly abnormal concentrations ($\Theta \neq 1$). Their characteristics are given by the Table **1.8**. (For damped neurons $U_C = U_{ON} = U_{OFF}$)

Table 1.8. Functional parameters of two typical damped neurons.

	p_1	p_2	Θ	f_1	f_2	U_{Thr}	H_{ON}
Swift	16.85	1.07	0.44	6.76	4.05	3.32	4
Slow	2.67	2.59	8.2	0.37	0.10	7.9	16

The initial F_0 firing rate ranges from 0 to 1500 s^{-1} for this swift neuron (in another example fitting the hearing nerve, the firing rate varies from 0 to 1000). Swift neurons seem to always have transient properties. When submitted to a constant excitation, the firing rate decreases quickly toward an asymptotic value. In our example, the asymptotic firing rate F_∞ is about 200 s^{-1}. It is reached in less than 0.1 s. for the other. Note the infinite slope dF/dU near the threshold.

The firing rate of slow neuron ranges from 0 to 60 s^{-1}. Transient neurons fit phasic motoneurons, non transient slow neurons the tonic ones.

Fig. (**1.20**) shows the initial F_0 and final F_∞ firing rate for swift and slow transient neurons.

1.3.3.3. Usefulness of the Various Typical Neurons

We use for this discussion a) the firing rate curves; b) the values N_1 of the synaptic strength index computed for a cell body radius $R = 3\mu$ (Appendix F); the experimental R values: from 1 to 50μ (1μ for little inter-neurons, 2μ for a cerebellum granule cell, 15μ for a cerebellum Purkinje cell).

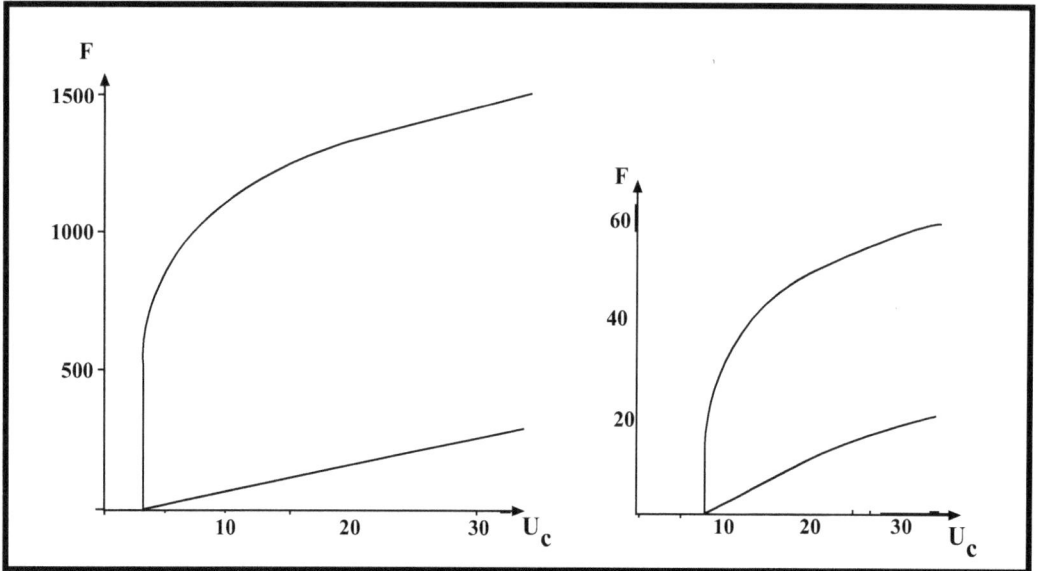

Fig. (1.20). Initial and final firing rate for typical swift and a slow neurons.

The *oscillating neurons* are either silent (F = 0) or firing with a rate in a narrow range (about 200 Hz). For R = 3μ, the coupling index N_1 as defined above is about 0.2. So, they are very convenient for a yes-or-no working with only one or two exciting synapses (with a minimum of one exciting synapse).

The *transient damped neurons* give first a short high frequency response, and then have a continuous firing rate proportional to U_C. To be combined with a linear synaptic excitation, they must be able to use a number of excited synapses varying from 1 to 20 (see paragraph III.2.3): U_C must be greater than U_{Thr} for only 1 excited synapse. So, when all the N synapses are excited, N > 20 N1. For R = 3μ, N_1 is about 1.6. For a Purkinje cell (R = 15μ), N has to be greater than 250. So, transient big swift neurons are convenient to combine linearly an important number of signals.

1.3.4. Systems Making Elementary Operations

Main Results: Logical and quantitative operations, threshold, competitive modulations, pace makers.

Introduction: We look systems built with a little number of neurons, some inputs and one (or several) outputs. The firing rate F_{Out} of trhe output depends on the firing rates F_1, F_2... of the inputs. The relation between F_{Out} and F_1, F_2... is an <u>elementary</u> <u>operation</u>. For instance, an addition is such that $F_{Out} = F_1 + F_2$.

The main idea is that the nature of an operation depends on a convenient choice of the neurons and synapses parameters. (So, neuronal processing cannot be understood without a previous study of the parameters). We present hereafter the simplest and the most useful devices. In each case, our assertions have been verified by a direct computation.

1.3.4.1. Logical (Boolean) Operations

1.3.4.1.1. Operation OR

An aperiodic neuron is excited by two yes-or-no synapses, each of them supplying an individual share to the control potential greater than the threshold U_{Thr} (for instance 1.4 U_{Thr}). The output fire if one OR the other neuron (or both) is excited.

1.3.4.1.2. Operation AND

Same share, but with an individual share 0.7 U_{Thr}. The output fires if the first AND the second neurons are firing.

1.3.4.1.3. Operation AND+OR

Same scheme as in the AND system, but with three inputs. The output fires if at least two of the three inputs are firing (if [I_1 AND I_2] OR [I_1 AND I_3] OR [I_2 AND I_3]).

1.3.4.1.4. Operation NEGATIVE

A spontaneous is inhibited by an input neuron. The output fires if and only if the input is silent.

Remind that, from Boole's theorem, any logical order can be obtained by a set of OR, AND and NEGATIVE operations.

1.3.4.2. Quantitative Operations

1.3.4.2.1. Linear Combinations

A swift transient neuron (linear F *vs* U_C relation) is excited by two linear input neurons through linear synapses (firing rates F_1 and F_2) and by a <u>modulating neuron</u> (constant firing rate F_M such as its individual share is exactly U_{Thr}). The, the output $F_{Out} = a.F_1 + b.F_2$ where a and b are two constants depending on the synaptic coupling strengths. (In fact, the modulation results from a feedback as we will see below). Without a threshold modulation, the system would be exceedingly insensitive to weak inputs.

Remark: Neuronal devices compute easily additions. But there is no simple neuronal device able to compute a multiplication or an exponential.

1.3.4.2.2. Transient Neurons and Time Derivative

The response of transient neurons is roughly the time derivative of the input U_C. Look at the response to a step function of the excitation U_C. The efficient excitation U_{Eff} exhibits an almost exponential decay which induces a progressive widening of the delay between spikes.

1.3.4.2.3. Time Integration

The activator concentration in a slow synapse is the time integration of the input firing rate during T_A seconds (or more exactly the convolution of F with a time constant T_A).

1.3.4.2.4. Slowly Increasing Signals, Impatience

It is made of a periodic neuron, a slow linear synapse and a linearly responding swift transient neuron. A short impulse triggers the firing of the periodic neuron; then, the activator concentration and the firing of the second neuron increase linearly during T_A (some seconds).

Slowly increasing signals (which we call impatiences) play a main part in all the searching devices.

1.3.4.3. Divergent Pathways

One input excites several divergent output pathways.

1.3.4.3.1. Progressive Recruitment

All the outputs have the same threshold, but the synaptic coupling (the number of synapses) with the input are different: $N_1 > N_2 > \dots$ Then, the input firing rate $F_{(1)}$ inducing the firing of the first output is weaker than $F_{(2)}$ which is weaker than $F_{(3)}$: the number of recruited output neurons increases when the common input excitation increases. Progressive recruitment is used to obtain logarithmic or exponential continuous signals, for instance to govern the force of the skeletal muscles.

1.3.4.3.2. Inverter Interneuron

We have seen that all the end synapses of a neuron are either excitatory or inhibitory. An inverting interneurons (Inv in Fig. **1.21**) is used to obtain however an exciting and one inhibitory pathways. This interneuron carries out a NEGATIVE operation.

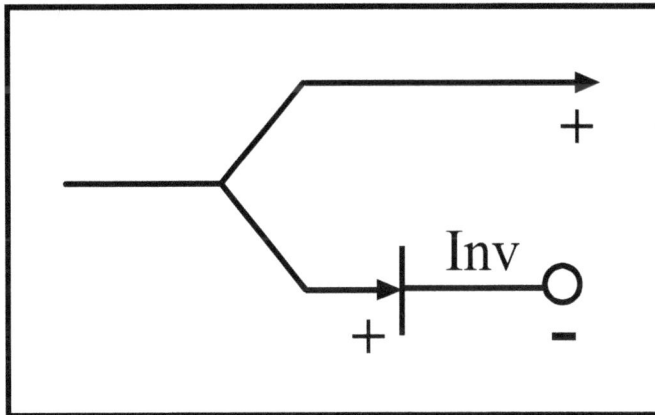

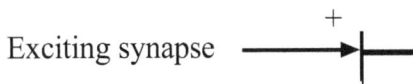

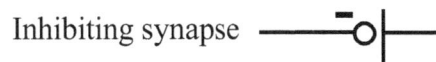

Exciting synapse ⟶+⊣   Inhibiting synapse ⟶−o⊣

Fig. (1.21). Scheme of an inverter system.

1.3.4.4. Modulations, Local Feedbacks

1.3.4.4.1. Operation IF and Threshold Modulation

We have seen that some operations (for instance a linear combination) need a threshold modulation to be carried out. When the modulation is silent, the output is silent. So, the device computes some combination of the quantitative (significant) inputs IF the modulation is firing.

1.3.4.4.2. Threshold Regulation

The regulated swift transient neuron excites an output pathway and, in the same time, is used to maintain its control potential just above the threshold. This loop induces a minimum firing rate close to 10 s^{-1}. The time constant of the loop is about 1 s (Appendix G).

1.3.4.4.3. Competitive Inhibition in Choosing Devices

The system is a set of several parallel aperiodic neurons, all alike. Each of them excites an output pathway and is excited by a linear input (or by a set of progressively recruited inputs). A common feedback is excited by the full set of outputs and inhibits in the same way all the parallel neurons. The result is (Appendix G) that the more intensively excited input inhibits all the others: only one output is firing at a time. Competitive inhibitions play a major part in the choice of behaviors.

1.3.4.5. Pace Makers, Delayed Signals

Two basic phenomena can be used: slow synapses time constant (some seconds), persistent firing of spontaneous or periodic neurons.

1.3.4.5.1. One Hour Memory

A short recording excitation triggers a periodic neuron. A short recording and stopping signal ends its firing. Inhibitions due to the sleep act as a security ending signal.

1.3.4.5.2. Heart Pace Maker

The heart pace-maker of arthropods [15] is a spontaneous neuron exciting the heart motor cells. Its firing rate is modulated by exciting and inhibitory inputs. In mammalians, the heart pace-maker is a double device made of low firing rate spontaneous cells. It uses the trend, for an oscillator, to be synchronized by weak impulsive periodic excitations with a slightly higher firing rate [16]. (This property results from the HH's equations, see Appendix D). The heart scheme described in the Appendix H interprets the normal heart working and some of its rhythm troubles.

1.3.4.5.3. The Pneumotaxic System [17]

Several master and slave oscillators control the breathing system. The upper one is called the pneumotaxic system. Located in the pons, it is made of 10^5 neurons. It is made of two sets of spontaneous neurons (controlling breathing in and out) and of a feedback using slow synapses to generate relaxation oscillations (Appendix I).

1.3.4.5.4. Detection of Sequential Signals

The echo locating device of the bat requires the measurement of the delay between emitted and received signals. More generally, a big lot of neurons responding to a sequence of two signals separated by Δt have been found in the auditory cortex of cats and monkeys. Such neurons play a great part in recognition of tunes and phonemes, in visual analysis of motion, and in imitative behaviors. A device using slow synapses (Appendix J) carries out an AND operation between two inputs occurring at time t_1 and t_2 if $\Delta_{Min} < t_2 - t_1 < \Delta_{Max}$. Usual order of magnitude for both Δ_{Min} and Δ_{Max} are 0.03 s $< \Delta <$ 3 s.

A slightly more complicated device generates repetitive signals: if triggered by a short input, it gives out a short output every $2T_A$ (about 10 s).

1.3.4.6. Circadian Devices

Circadian neurons form a master clock which governs a lot of secondary clocks [18]. Some of them could be supported by non neural cells: a secondary clock has been located in the liver.

A direct line connects the eye to the suprachiasmatic nuclei. As any oscillators, the circadian neurons are synchronized by periodic signals with a frequency slightly greater than the spontaneous frequency. Thus, in man, the hormones concentrations obey to a daily frequency. More complicated devices exhibit a daily frequency. For instance, the activity of diurnal mammals in very hot countries has two maxima, one the morning, the other in the evening. We know daily alarm clocks which measure the time elapsed since the sun set, for instance giving a signal at meal time. We know also chronometric drivers: a cat needs to watch about one hour daily, to catch about ten minutes; without mouse, it plays with a curtain. Short cycles are known, for instance the T = 90 minutes duration cycle of rapid eyes motions (REM) during sleep.

1.3.4.7. Random Time Generators (Operation "Wait for a While")

In light, nocturnal mosquitoes are motionless. Darkness is established at time t=0, the mosquitoes wait some time before flying away. Measuring (for several mosquitoes and several times for the same mosquito) the time t of the flight away, one finds a very simple probabilistic result [19] with a mean time about 20 minutes. The device uses probably the waiting of an infrequent event: the perfect coincidence of the spikes of several slightly different periodic neurons (see Appendix K).

1.3.5. General Aspects of Huge Neuronal Sets

Main result: Functional limitations due to the neuronal noise, massive parallelism, hierarchical modulations.

1.3.5.1. The Neuronal Noise

The performances of any transmission line are limited by the bandwidth and by the noise. We want to apply this general result to nerves. The bandwidth is clearly related to the range of possible firing rates. But what is the neuronal noise?

Among several candidates (for instance, random release of transmitter), we assume that the main source of neuronal noise is the electric interaction between close and independent cell bodies. In HH's theory, one assumes an infinite external medium; in fact, neurons form tight bundle, lines of electric current from one neuron reach other neurons and so perturb the value of the potential (Fig. **1.22**): the control potential of each neuron is perturbed by the spikes of other neurons which act as random noise sources.

Assuming that the parameters of this noise are the same for all the neuronal transmission lines, we had to evaluate it and found that the noise is a random component (mean value 0, width at $2\sigma \approx \pm 0.15$ mV) of the control potential. (Note that 0.15 mV is not a magical number, but the order of magnitude of a number depending on the sub-unit we look at and ranging probably from 0.1 to 0.25 mV).

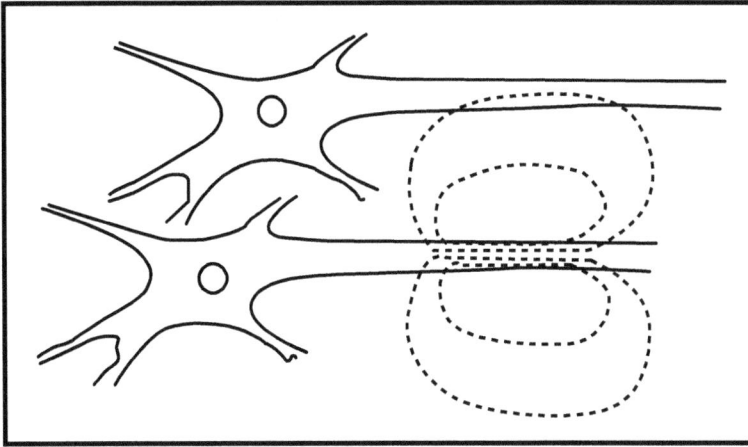

Fig. (1.22). Electric perturbation of the control potential of a neuron by a neighbor.

This assumption is supported by the interpretation of several experimental results: a fairly direct measurement (Appendix L); the analysis of threshold regulation and of α EEG waves (Appendix M); the trick used in retina to suppress noise when detecting very weak luminous signals and we will see above that the assumption leads to reasonable correlations between the anatomy and the performances of some neuronal devices.

1.3.5.1.1. Analysis of a Direct Example

We started from the experimental study [20] of spontaneous firing of ganglion cells in cat retina (ganglion cells are the third layer of the retina; the bundle made of their axons is the optic nerve). Each of these cells is excited and inhibited in a complicated way by many photoreceptors but many of them are silent in absolute darkness: to obtain clear results, we analyse their spontaneous activity without external excitation, it is to say in absolute darkness. The experimental result is the statistic of the delays T between two successive spontaneous spikes. This statistic is somewhat strange since it emphasis large values of T against small values (exhibiting some very great values of T, but no very weak value). This experimental result does fit quite well with the theory.

1.3.5.1.2. How Retina Escape to Noise

In diurnal vision, the weakest detectable illumination generates control potentials about 0.15 mV in cones, bipolar cells and ganglions cells, an adjustment, which is compatible with our proximity noise.

On the contrary, in nocturnal vision, the weakest detectable illumination generates very weak control potentials reaching 3.10^{-4}, 6.10^{-3} and 0.15 mV in rods, bipolar cells and ganglion cells.

Another experimental fact is revealed under a very weak illumination: the existence of lateral gap junctions. When the illumination is strong, rods are independent. But, lateral diffusion of the excitation (that is lateral ionic exchanges) has been measured in weak illuminations. Lateral diffusion spreads till 200 μ in tortoise and 50 μ in man [15]. As the rod diameter is 8μ in tortoise and 2.5 μ in man, lateral diffusion causes the coupling of about $(200/8)^2 = 625$ or $(50/2.5)^2 = 400$ rods in tortoise or man and decreases by a factor $\sqrt{400} = 20$ the angular resolution.

If we assume a synaptic noise, the lateral diffusion of excitation would not decrease the noise level. If we assume a proximity noise, the random character of which is due to the independence of weakly interacting neurons, the lateral diffusion of excitation causes the coupling of adjacent cells, increases the size of the receptive field and suppresses the noise. Thus, to reach sensitivity to very weak excitations, angular keenness is sacrificed to sensitivity.

1.3.5.2. Number of Parallel Neurons and Global Performances

How many different messages can be transmitted by a neuron? (This number is the transposition of the transmitting capacity of a line). For high frequency transient neurons (phasic neurons), the number (Nbr) depends on the transient time constant T_D. For slow linear devices (tonic neurons), the capacity is noise limited.

1.3.5.2.1. Case of a Tonic Sensorial Line: The Ruffini's Knee Sensors

A neuron starts firing if $32 > U_C > U_{ON}$ with U_{ON} close to U_{Thr} and nearly neglectible. The control potential is read with an approximate accuracy $\delta U = 0.15$ mV (two times the standard deviation of the noise). Then the neuron is able to differentiate between Nbr = 32 / 0.15 different messages. The order of magnitude of Nbr is very low: about 200. The time constant Tc of the receiver neuron should be such that $T_C F_{MAX} \geq$ Nbr.

The Ruffini's sensors measure the angle between femur and tibia. The range is from 50 to 180 degrees. We detect with an accuracy of about 1 mm the height of our foot above the earth, which means an accuracy of 0.1 degree. Thus, the sensor detects 1300 different possible messages.

Ruffini's corpuscle is a typical aperiodic neuron (with a special input synapse). It can distinguish 200 different excitations. The sensorial nerve has to be made of 1300/200 (equal roughly to 7) neurons. This result fits with the anatomical findings.

1.3.5.2.2. Case of a Phasic Sensorial Line: The Auditory Nerve

From physiology, we know that the ear measures a two components quantity: one is the logarithm of sound frequencies in the range 16 to 2.10^4 s^{-1} with a discrimination threshold $\Delta F / F \approx 3.10^{-3}$. The second component is the sound intensity with a range about 120 dB and a discrimination about 1 dB. The first component can take 2400 values, the second 120. The number of different possible messages is then $2400 * 120 = 3.10^5$.

From anatomy, we know that the internal hair cells (the sensors) excite 27000 bipolar cells (the auditory nerve is made of their axons). The bipolar cells are typical swift neurons with a very short decreasing time (T_D about 10^{-2} s) and a firing rate ranging from 0 to 1500. Then each cell can only transmit F.T ≈ 15 different messages and the nerve would be made of 2.10^4 cells (when the direct experimental result is $2.7\ 10^4$).

1.3.5.2.3. Case of the Command of a Skeletal Muscle: The Biceps Femoris Muscle

Made of 450 fibres (tonic and phasic), it governs the knee position. We have seen that the knee can take 1300 different positions.

The tonic muscular fibres are imbedded in a spinal regulating device. The reading time of the signal cannot be greater than the time constant τ_R of the regulation (about 0.2 s for a tonic automatism). The fibres are excited by typical slow neurons without transient properties. Their firing rate range is 0 to 60 s^{-1}. The threshold voltage of the exciting neurons is around 8 mV and the maximum value of their control voltage Uc is 32 mV. Therefore the number of possible messages that an exciting neuron can transmit is 24/0.15 = 160 and a total of 8 neurons is sufficient to code the 1300 positions of the knee. 1300 positions are possible to code on 10.34 bits. If these 10.34 bits have to be passed in 0.2 seconds, it can be shown that the efficiency of the transmission channel between the neurons and the tonic fibres roughly is 60% (Appendix N).

There is a limitation T of the reading time (for instance 0.03 s for the retinal persistence). A first limitation is due to the transient time constant. Very often, interneurons break off more frankly the reading: lateral geniculate body cuts the visual reading each 0.03 s while the eye is moving, Renshaw's neuron cuts the firing of phasic motoneurons each 0.05 s. Then, if the neuron firing rate ranges between F_{Min} and F_{Max}, the neuron is able to transmit $(\Delta F = F_{Max}-F_{Min}).T$ different messages. In this example: Fmax = 60, Fmin = 0, T = 0.05, Number of messages per phasic neurons: 3; Number of phasic neurons: $1300/3 \sim 433$. Computed total number of neurons: 433 phasic neurons + 8 tonic neurons = 441 neurons. This computed value has to be compared with the anatomic appraisal of 450.

1.3.5.2.4. Other Example: The External Right Ocular Muscle

This striated but non skeletal muscle governs the eye orientation. The range of eye orientation is 40 degree with accuracy of 6": it can distinguish 23000 different positions. The eye chooses an orientation every T = 0.03 s.

The motoneurons are a variant of slow neurons with $\Theta = 2.7$ in place of 8.2 with an F range is [0 - 180 s^{-1}]: during T, they can transmit 5 different messages. The nerve would need 4600 neurons, a number compatible with anatomy.

1.3.5.3. Massive Parallelism and Cortical Architecture

1.3.5.3.1. Massive Parallelism

The limitation of the transmitting capacity of neurons requires a multiplexing coding of the sensorial information and the use of many parallel neurons: they are 10^6 parallel in the optical nerve. The first visual cortical area is a set of sub-units made of a great number of identical columns (about as many columns as individual receptors). Each one is build of a great pyramidal cell and ten or more little interconnecting neurons. The anatomical arrangement of the columns in a sub-unit is strongly correlated to their functional identification [21] and amazingly systematic (Fig. **1.23**). In the visual area 17, long pyramidal neurons and their columns are distributed among "paving stones" (coordinates X, Y) which map the retina geometry (coordinates x, y). Each paving stone is associated with one value of x, one value of y and with the dominance of one or the other eye. A horizontal motion dx on the retina is associated with a displacement along the X axis of the area. Ocular dominance zones alternate along the X axis. Inside each paving stone, a vertical motion on the retina is associated with a Y displacement on the cortex. This area is sensitive to luminous bars. Inside each paving stone, circular displacements as represented in the figure are associated with the angular orientations θ of the luminous bar.

The systematic and modular character of the circuitry seems a general property of the cortex.

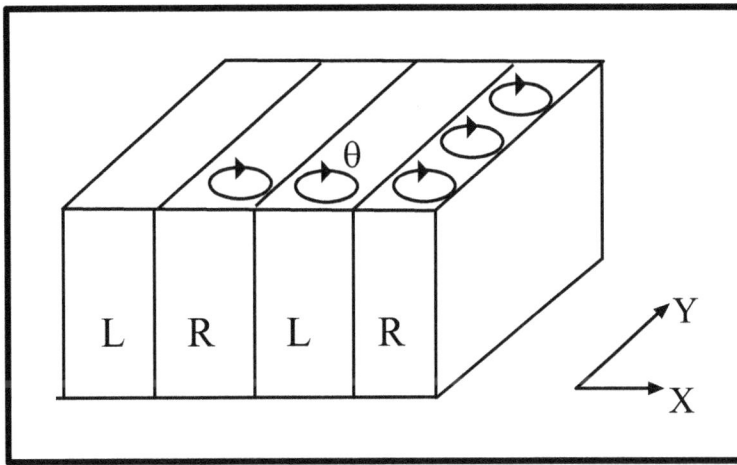

Fig. (1.23). Architcture of the visual cortex.

1.3.5.3.2. Histological Layers and Functional Stages

In the cortex, the signals flow through a several floors of sub-units, each one located beside to the others. Thus, the human smell analyzer is made of 6 floors of 10^4 columns. The anatomical layers of the cortex correspond to this organization. The bodies of the pyramidal cells form vertical (perpendicular to the brain surface) columns. Input and output axons form horizontal layers connecting one cortical sub-unit to the others and to sub-cortical ones (Fig. **1.24**).

Layer I (at the surface of the brain) connects cortical sub-units. Layer IV is made of the inputs coming from the thalamus (and especially from the geniculus). Layer VI is made of the outputs toward the thalamus.

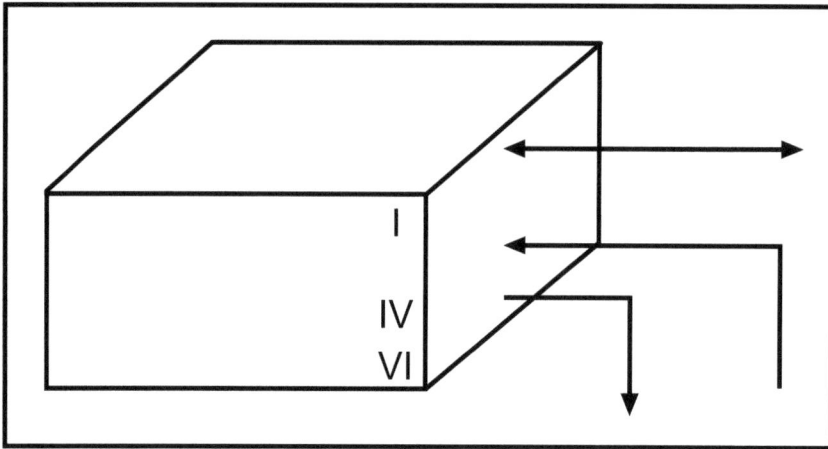

Fig. (1.24). Connections between layers of the cortex.

1.3.5.4. Modulations

Some neurons (located inside a sub-unit or in other parts of the CNS) excite or inhibit all the columns of a sub-unit. The modulating signal is transmitted by several modulating synapses or by many common synapses.

1.3.5.4.1. Hierarchical Modulations

In a very simplistic way, we regard the encephalon as a pile of three strata. The deeper is sometimes called "paleo-mammalian brain". It is made of the hypothalamus, the septum, the amygdalae and the hippocampus. It directly controls the emotions, the neuro-endocrine interactions, the main behaviors (sleep or wakefulness, nourishment, defence, sexual activity) and would play an important part in memorizing process. Stimulations of well-chosen places in it induce sham rage or hyper sexuality in cats or transform a furious bull into a pacific animal. The second stratum is made of the thalamus and the grey motor nuclei. The third stratum is the neocortex. The deeper makes rough analysis of signals. It exerts modulations on the second stratum, which proceeds to a more detailed analysis and exerts modulations on the cortex, in charge of the finest analysis.

1.3.5.4.2. Retrograde Regulations

Looking at a normal flow of signals from upward (the sensors) to downward (the muscles), many modulations appear as retrograde: they start from downward to excite (or inhibit) upward sub-units. Retrograde modulations modify the processing of signals following the first results (the interest, the meaning) of this analysis. We will see in part IV how retrograde modulations govern the recording and the reading functions of memorizing devices.

1.3.5.4.3. Miscellaneous

We have often to deal with signals modulating only a part of a sub-unit. Superposition of several partial and weak modulations leads to a great flexibility in choices. For instance, such superposition of modulations plays a major part in language control.

Some modulations play a local logistic part (threshold regulations, security devices).

Hierarchical inhibitions triggered by a particular neuron of the controlled sub-unit act as Freudian taboos.

1.3.5.5. Transmission *Versus* Selection

In retina, the number of neurons in each level and the operations between two stages are such that the information is coded and transmitted without significant losses. This is an example of transmitting sub-units.

On the contrary, we have described (Section 1.3.4.4) a circuitry with inhibiting competitive feed-back: the more intensively excited neuron (or column) fires and forbids the firing of all the others. Such a sub-unit does not transmit information, but select one pathway among several incompatible possibilities. We will see (Part III) that competitive inhibition plays a main part in sensorial analysis.

APPENDICES TO THE SECTION 1.3

A: Pulsation ω and damping f of little oscillation

B: The threshold potential U_{Thr}

C: Response to a constant excitation

D: Response to impulses

E: The synaptic coupling strength

F: Firing rate and synaptic strength index for typical neurons

G: Local feedbacks

H: A rough scheme of the heart pace-maker

I: The pneumotaxic system

J: The detection of sequential signals

K: Random time generators

L: The spontaneous random firing of a cat ganglion cell

M: α EEG waves

Appendix A: Pulsation ω and Damping f of Little Oscillation

Little oscillations obey to the driving equation:

$$\partial^2 U/\partial\, t^2 + 2f\, \partial U/\partial t + (f^2 + \omega^2)\, U = (f^2 + \omega^2)\, U_C$$

The damping and the pulsation are given by a second degree equation with two complex roots $f \pm i\, \omega$ or with two real roots f_1 and f_2. Little oscillations are then the sum of two negative exponential $\exp(-f_1 t)$ and $\exp(-f_2 t)$. Their values are:

$$f_1 + f_2 \text{ or } 2f = 2p_1\,(0.355p_2 - 0.245) + 0.2$$

$$f_1.f_2 \text{ or } f^2 + \omega^2 = p_1\,(0.382p_2 - 0.09)$$

For slow and swift neurons, the above values of f and ω must be divided by Θ.

Hints at the Computation

We use Laplace transform to study first order perturbations of the HH equations around the rest conditions $U = 0$, $\partial U/\partial t = 0$ and $\partial U/\partial x = 0$. To do so, we expanse each function (I, $\partial I/\partial x$, meq, τ_m) in powers of U and look only to the first term. We make a crude approximation by taking for the fast opening m gate of the Na channel at rest $\tau_m = 0$ and for the two others $\tau_h = \tau_n = 5$ seconds. We obtain thus a second order equation in s and we compute the eigen values f and ω.

If Θ is not one, we have to multiply the time scale by Θ.

Appendix B: The Threshold Potential U_{Thr}

One found:

$U_{Thr} = (3p_2 + 0.1) / \{1 - \exp[-p_1(0.71p_2 + 0.02)]\}$

Note that the threshold does not depend on the time scale factor Θ.

Hints at the Computation

*The beginning of the spike is due to the opening of the fast gates of Na channels (m factor) while the slow gates (n and h factors) keep their rest value. In the Hodgkin Huxley equations, we obtain a useful approximation by taking for n and h their values n_0 and h_0 for the rest potential $U = 0$, by taking $\tau_m = 0$ (which implies that m = meq) and for **meq**(U) the first two terms of the expansion in powers of U. Reporting in the HH equation for $I = 0$ and naming A and B the constants, we find a Bernouilly equation, the solution of which is if $U = U_0$ at time $t = 0$:*

$$U = \frac{AU_0 \exp(-ABt)}{A - U_0[1 - \exp(-ABt)]}$$

We assume that a spike is triggered if U becomes infinite before $t = 1$ ms. Thus:

$$U_{Thr} = \frac{A}{1 - \exp(-ABt)}$$

Appendix C: Response to a Constant Excitation

Neglecting any x dependence, we integrate the driving equation after precising the initial conditions and with a constant value of U_C.

a)   *First spike of a neuron at rest:* With U_C slowly increasing, $U = U_C$ before the spike. Then the neuron fires when $U_C = U_{Thr}$

b)   *After a first spike:* the end of spike conditions are $U = -10$ mV and $dU/dt = 0$. For oscillating neurons, the solution is:

$$U = U_C - (U_C + 10) \exp(-ft) [\cos(\omega t) + (f/\omega) \sin(\omega t)]$$

The neuron fires when $U = U_{Thr}$ If this occurs at time t and if θ is the duration of the spike, then the firing rate F is $F = 1000/(t + \theta)$.

The maximum value of U occurs when $\omega t = \pi$. Then U_{OFF} is the value of U_C such that $U = U_{Thr}$ for this time.

c)   *ON conditions:* A neuron at rest is submitted to a step-like excitation (in place of a slowly increasing excitation). The end initial conditions are $U = 0$ (in place of-10 mV) and $dU/dt = 0$. Then:

$$U = U_C \{1 - \exp(-ft) [\cos(\omega t) + (f/\omega) \sin(\omega t)]\}$$

The maximum value of U occurs when $\omega t = \pi$. Then U_{ON} is the value of U_C such that $U = U_{Thr}$ for this time.

d)   *Damped neurons:* We obtain easily similar formulas for real eigen values f_1 and f_2, for instance after a spike:

$$U = U_C - (U_C + 10) [(f_1 \exp(-f_2 t) - f_2 \exp(-f_1 t)] / (f_1 - f_2)$$

U(t) is an always increasing function of time. Then $U_{OFF} = U_{ON} = U_{Thr}$

Appendix D: Response to Impulses

The impulse is simulated by a Dirac's function. For an oscillating neuron at rest, we find after an impulse with height H:

$$U(t) = H\Theta [(f^2 + \omega^2)/\omega] \exp(-ft) \sin(\omega t)$$

Its maximum occurs at the time $t_{Max} = (1/\omega) \text{Arctg}(\omega/f)$

Then, H_{ON} *is such that* $U(t_{Max}) = U_{Thr}$

Note that we can compute the response to an impulse plus a constant excitation or to 2 impulses or to any complicated situation.

Trends Towards Synchronization

After a first excitation by a spike at t = 0, a periodic neuron is is excited by another impulse at time t_{imp}. Let T = 1/F the interval between two successive spikes. Without impulse, T would have the value T_0. Using the above formulas, we compute T as a function of t_{imp} (Fig. **1.25**).

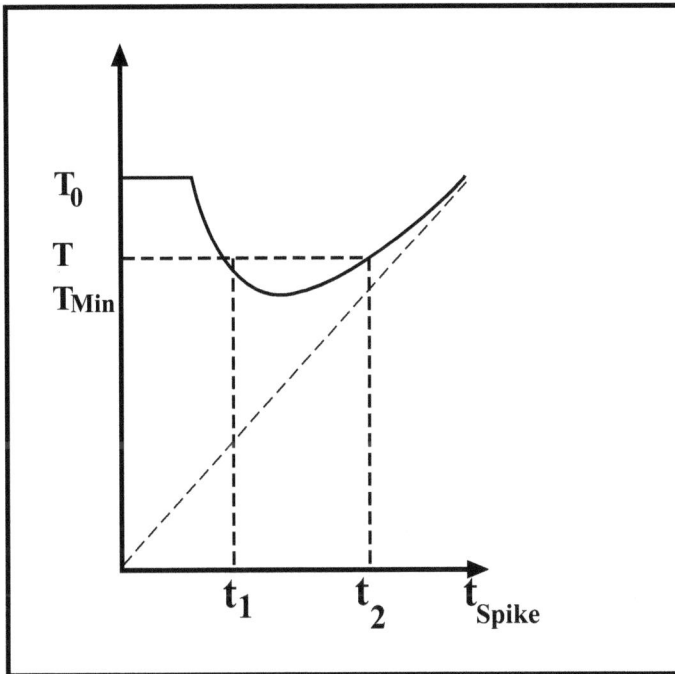

Fig. (1.25). Synchronisation of a periodic or spontaneous neuron by periodic excitations.

If $0 < t_{imp} < \theta_{spike}$ the impulse does not act and T = T_0. For $\theta_{spike} < t < T_0$, the curve $T(t_{imp})$ reaches a minimum T_{Min}.

We look now to a periodic impulsive excitation occurring at times t_{imp}, $t_{imp} + T$, $t_{imp} + 2T$...

If $T_{Min} < T < T_0$ we find on the curve two solutions at times t_1 and t_2 One sees easily that t_1 is an unstable solution and t_2 a stable one: periodic impulsions may synchronise a periodic neuron on a firing rate F = 1/T slightly greater than the free firing rate F_0 Then, the driving impulsion occurs just before the beginning of the spike.

This discussion can be extended to any kind of neuron.

Appendix E: The Synaptic Coupling Strength

Looking to the addition of excitations in a case without inhibiting synaptic signals and with strong exciting signals (then $\Omega = 1$), we find:

$$U_{Thr} = 32 \frac{N}{N+K} \ or \ N_1 = K\left(\frac{U_{Thr}}{32 - U_{Thr}}\right)$$

Appendix F: Firing Rate and Synaptic Strength Index for Typical Neurons

Firing Rate Versus the Control Potential (Tables 1.9-1.14)

Table 1.9. F *versus* U_c law for an aperiodic neuron.

U_C	2.63	2.79	5.77	11.64	20.6	32.1
F	140	150	200	250	300	345

Table 1.10. F *versus* U_c law for a periodic neuron.

U_C	-1.17	-0.81	1.56	5.77	12.12	21.35	32.2
F	176	200	250	300	350	400	442

Table 1.11. F *versus* U_c law for a spontaneous neuron.

U_C	-5.55	-5.03	-3.98	-0.58	4.82	12.95	24.95	31.9
F	146.7	175	200	250	300	350	400	422

A rough fit of the $U_C = \Phi(F)$ relation is $\Phi = a + b (F/100) + c (F/100)^2$ with:

Table 1.12. Parameters for the fitting law.

	a	*b*	*c*
Aperiodic	*13.80*	*-17.03*	*6.5*
Periodic	*14.87*	*-17.73*	*4.9*
Spontaneous	*9.17*	*-18.26*	*5.6*

This function Φ is not defined if $U_C < U_{OFF}$

Table 1.13. F *versus* U_c law for a swift neuron.

U_C	3.32	$3.32+2.10^{-5}$	3.38	4.23	7.24	14.45	31.7
F	0	250	500	750	1000	1250	1500

Note the brisk onset of the response

Table 1.14. F *versus* U_c law for a slow neuron.

U_C	7.90	7.9025	8.3	10.1	14.0	20.7	31.8
F	0	10	20	30	40	50	60

The $U = \Phi (F)$ curves of these two neurons are fitted by:

Swift neuron: $\Phi(F) = U_{Thr} + (F/\Psi)^n$ with $U_{Thr} = 3.32$ mV, $\Psi = 760.6$ s^{-1} and $n = 5$

Slow neuron: $\Phi(F) = U_{Thr} + (F/\Psi)^n$ with $U_{Thr} = 7.93$ mV, $\Psi = 24$ s^{-1} and $n = 3.5$

Note that $U(F)$ is defined for every value $F > 0$ and that the slope dF/dU is infinite near the threshold: $d\left(U - U_{Thr}\right) = \left(\dfrac{n}{\Psi}\right)\left(\dfrac{F}{\Psi}\right)^{n-1} dF$ with $n>1$ and $F=0$

Coupling Strength Index N_1

K is proportional to $R^{3/2}$. For the standard value $R = 3\mu$, we obtain the Table **1.15**.

Table 1.15. Typical values of the coupling index.

	Aperiodic	Periodic	Spontaneous	Swift	Slow
K	1.4	1.8	1.4	13.6	0.5
$U_{threshold}$	6.0	3.2	2.4	3.3	7.9
N_1	0.3	0.2	0.1	1.6	0.2

Appendix G: Local Feedbacks

Threshold Regulation of a Swift Transient Neuron

The control potential of the regulated neuron is the sum of the contributions of 2 inputs: a) a significant signal (firing rate F_S) carried by a linear pathway; b) a modulating neuron (firing rate F_M). The control potential of the regulated neuron is $U = U_S + U_M$.

The modulating neuron is a spontaneous one. When it fires freely, its contribution U_M is definitely greater than the threshold U_{Thr}, for instance $Max(U_M) = U_{Thr} + 1$ mV The modulating neuron can be inhibited; the strongest inhibition does not stop the neuron, but leads to a decrease of U_M which becomes slightly smaller than the threshold, for instance $Min(U_M) = U_{Thr} - 0.3$ mV.

A feedback links the output of the regulated neuron to the modulating neuron: The output of the regulated neuron excites a yes-or-no feedback neuron through very strongly coupling synapses. So, the feedback firing rate $F_{FB} = 0$ if the regulated neuron is silent and F_{FB} takes its maximum value as soon as the firing rate of the regulated neuron is greater than a very weak value (for instance $F_{Out} = 10$ s^{-1}).

The feedback neuron inhibits the modulating one through a slow synapse with T_A about 1 s. Without significant input, $U \approx U_{Thr} + 1$ mV, the device oscillates with a T_A period around a very weak mean output frequency. With a notable significant input U_S, the maximum inhibiting feedback is reached leading to a negligible input correction and $U \approx U_S + U_{Thr} - 0.3$ mV. Hence, the output firing rate is almost proportional to U_S. The α EEG waves result from the threshold regulation of cortical neurons: so, from the EEG records, we were able to evaluate the values of T_A and of the mean rest firing rate.

Competitive Inhibition in Choosing Devices

*Here, we deal with periodic neurons excited by independent linear inputs and inhibited by a common feedback neuron (Fig. **1.26**).*

One firing output is sufficient to obtain the maximum feedback firing rate. Slow synapses inhibit all the inputs. Thus, the control potential of the more strongly excited pathway takes a value greater than U_{ON} inducing the firing of this pathway while other pathways are utterly inhibited.

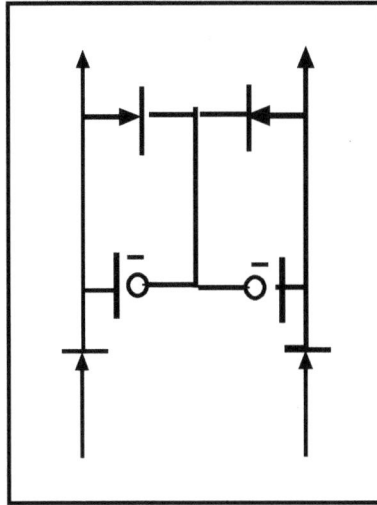

Fig. (1.26). Competitive feedbacks in a choosing device.

Appendix H: A Rough Scheme of the Heart Pace-Maker

Experimental Results

Cardiologists know that the control device is made of two oscillators linked by the Hiss bundle (Fig. **1.27**).

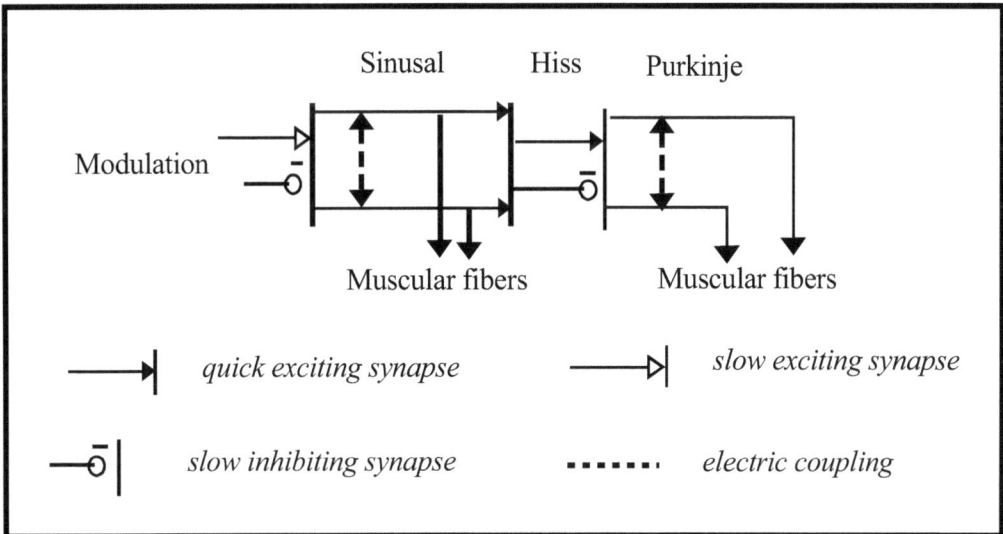

Fig. (1.27). The coupled cardiac pacemakers.

The Keith and Flack's sinusal node (the master oscillator) controls the contraction of atria and excites (through the slow conducting Hiss' bundle) the Purkinje's cells (the slave oscillator) which cover the ventricles and control their contraction. The occurrence of fibrillation proves that the muscular fibres are independent. Their synchronization is forced by the pace-maker.

Constraints to be satisfied

a)   Heart mechanics led to firing rates about $1 \, s^{-1}$.

b)   The fibres of the heart muscle are excited in accordance with a fix sequence in time. The delay between the excitation of each fibre is defined by a "delay line", in fact the length of an axon. The origin in time of the sequence of excitations would have to be synchronized to a time basis defined by a master oscillator. In fact, in order to reach a sufficient reliability, synchronization process calls for a number of redundant oscillators. The difficulty is to conciliate synchronization and independence of the oscillators.

The Excitable Cells (Sinusal Node and Purkinje's Cells)

Abnormal Ionic Concentrations

With a time scale factor $\Theta = 210$, the firing rate range of the analogue of a typical spontaneous neuron is 40 to 120 min^{-1} ($F_0 = 70 \, min^{-1}$ without any excitation) and the spike duration (1 ms multiplied by Θ) is $\theta_{spike} = 0.2 \, s$. The rest potential $V_R = -103 \, mV$ fits well the measured ionic concentrations.

The Sinusal Node

The inputs of the node are a) external modulating inputs transmitted through slow synapses. They modulate the firing rate F; b) feedback exciting inputs (probably due to electric gap junctions, each cell receiving a weak impulse from all the other firing cells. This retroaction forces the mass synchronization: the dispersion of the spikes of all the cells is less than 1 s. The outputs of the sinusal node excite some muscular fibres and (through the Hiss fasciculus) the ventricular node.

The Ventricular Node (Purkinje's Cells)

Each Purkinje's cell receives impulses delayed from a constant value for each spike of the sinusal node. b) Through some inverting relays and a slow synapse, a slowly changing inhibitory signal. The sum of the impulsive excitation and the continuous inhibition makes the Purkinje's cells to be synchronized on the sinusal node spikes. c) a weak internodes excitatory feedback coordinating the firing of the Purkinje's cells.

Some Possible Causes of Breakdown

If the inhibition is too weak, the slave regains its liberty; its firing rate differs from its master one: it is arrhythmia. If the sinusal node stops, the Purkinje's cells continue for a time to be inhibited without receiving any impulse: they stop. Then the concentration of activator in the inhibitory synapses slowly decreases, spontaneous firing of the Purkinje's cells occurs: it is the pre-automatic breakdown. If the firing rate of the sinusal node abruptly increases, the inhibition of the Purkinje's cells does not increase at first while their excitation increases, the slave sometimes fires twice after one spike of the master: we observe extra-systoles.

Appendix I: The Pneumotaxic System

Electro-physiology found in it three firing rate *versus* time behaviors (Fig. **1.28**).

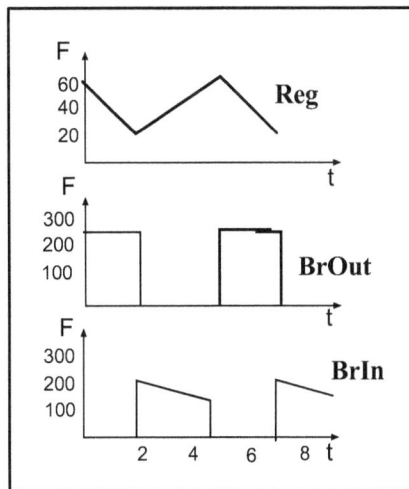

Fig. (1.28). The three firing rates observed in the pneumotaxic system.

BrIn and BrOut types govern the neurons activated during breathing in or breathing out and that the oscillating Reg type (Reg for regulating) govern the sequences of in and out breathing.

Interpretation

The device is an oscillating device regulated by an external feedback (Fig. **1.29**). *Neurons and synapses parameters: BrIn and BrOut are typical spontaneous neurons. The regulating neuron is a typical slow neuron. The BrIn / Reg synapse is a slow one with a weak $\gamma.T_A$: Ω increases slowly and linearly versus time when BrIn is firing, decreases when it is silent. Reg inhibits BrIn. When the firing rate of Reg is 60 s^{-1}, it induces in BrIn a negative control potential $U_C = U_{OFF}$ and the neuron stops. When the firing rate of Reg is 23 s^{-1}, BrIn begins to fire.*

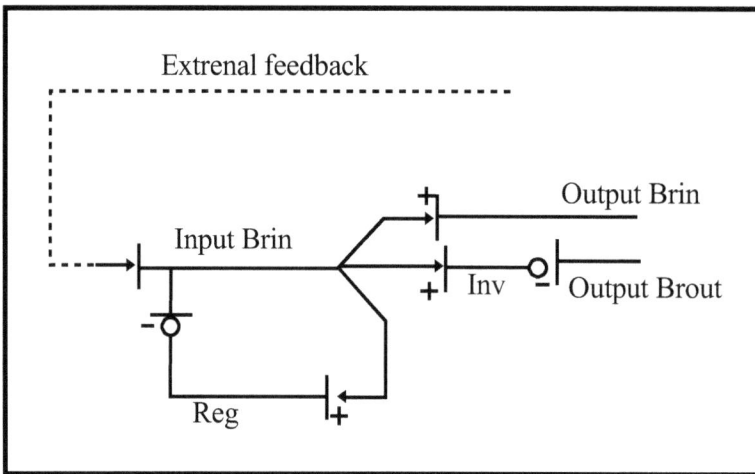

Fig. (1.29). Neurons of three kinds govern breathing.

In these conditions, the duration of the breathing in is 3 s, the duration of the breathing out 2 s. (These numbers change when BrIn receives modulating excitations). The BrOut output is the NEGATIVE of BrIn.

Appendix J: The Detection of Sequential Signals

The input neuron I_1 generates a short burst of spikes at time t_1. It excites through a swift synapse a periodic neuron P (Fig. **1.30**), which excites through a slow synapse a periodic neuron D.

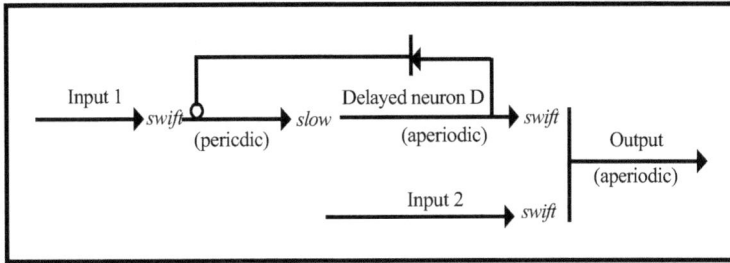

Fig. (1.30). A first signal is kept till the occurrence of a later signal.

Its threshold is reached after a delay Δ_{MIN}. An inhibiting feed-back stops the firing of P after an further delay $\Delta_{FEEDBACK}$ (and $\Delta_{MAX} = \Delta_{MIN} + \Delta_{FEEDBACK}$). Thus, the delayed neuron D fires from $t_1 + \Delta_{MIN}$ to $t_1 + \Delta_{MAX}$. Both the neuron D and the second input I_2 excite through swift synapses an aperiodic output neuron O. For convenient values of the synaptic couplings, the output control potential becomes greater than the threshold if and only if D and I_2 are simultaneously firing (operation AND).

Appendix K: Random Time Generators

The probability of the observed flying away time for mosquitoes is Prob $(t > \theta) =$ exp$(-\theta/T)$. The time constant T is the same for all observed mosquitoes: about 20 minutes

Coincidence of Spikes as a Random Process

Look at two neurons firing with two different periods T_1 and T_2. A spike of the second neuron occurs at time kT_2 (where k is some integer). The relative phase of this spike is $\varphi_k = kT_2$ modulo T_1. It is well known that, for great k, the phases are uniformly and randomly distributed on the segment $[0, T_1]$.

Scheme of a Convenient Circuitry

The circuitry is made of $(\zeta+1)$ slightly different periodic neurons; all of them excite the same swift neuron. *The firing rate of unexcited periodic neurons is about 230 s^{-1}. So, their periods T_p are about 4 ms. The damping coefficients f_1 and f_2 of the swift neuron are 7 and 4 ms^{-1}. Thus, an impulsive excitation is forgotten after about $\theta = 2/f_2 \approx 0.5$ ms $= 0.13 \, T_p$. The minimum impulsive excitation of the swift neuron is $H_{ON} = 4$ mV. We assume that the individual excitation H due to each periodic neuron is such that $\zeta.H < H_{ON} < (\zeta+1) H$. Then, the swift neuron will fire if the spikes of all the periodic neurons occur within a time less than θ. The probability p of such an*

*occurrence is $p = (\theta/T_p)^\zeta \approx 0.13^\zeta$ and the mean waiting time is $T_W = T_p/p$. For $T_W \approx$
20 minutes, $\zeta = 6$ periodic neurons.*

Appendix L: The Spontaneous Random Firing of a Cat Ganglion Cell

Experimental Results

The authors looked at the spontaneous firing of ganglion cells in the retina of the cat
in darkness. They studied the impulse interval T distribution [20].

The distributions adjusted either on one thousand samples of a hundred milliseconds
duration or on hundred samples of one second duration lead to a Gaussian probability
density function (pdf) with a mean value of 3.8 pulses by second and a standard
deviation of 2.1.

Interpretation

The ganglion cells are transient swift neurons. As no other excitation is involved,
noise is the only reason that may explain the firing of spikes in the ganglion cells of
the retina.

*Near the threshold, the equilibrium relation between the firing rate and the control
potential of such neurons is almost linear. We assume that, in darkness, the control
potential is:*

$U_C = [U_{Thr} + \delta U] + w$

*where the bracketed terms are due to a regulating system and where w is a random
Gaussian excitation. As the answer of the neuron is linear, we can write:*

$$\frac{F}{F_D} = \delta U + w \text{ and } w = \frac{F - F_D \, \partial U}{F_D}$$

*Assuming that w is a Gaussian random process with 0 mean and standard deviation
σ, we calculate the probability density function of F in relation with the δU one:*

$$f(F) = \frac{1}{\sigma F_D \sqrt{2\pi}} \exp\left[-\frac{\left(F - F_D \, \partial U\right)^2}{2\left(\sigma F_D\right)^2} \right]$$

*According to the experimental results summarized by the adjusted curve that reaches
its maximum value for f(0)= 0.192, $F_D \, \sigma$ = 2,1 and if F_D = 28 mV^{-1}, σ = 0.075 mV.*

The adjusted curve shows a maximum value for F ~ 3.8 and the calculated Gaussian distribution has a mean value F = 4.15

It is difficult to analytically deduce the distribution probability of the time intervals between two spurious spikes from the distribution of F. Generally, in renewal processes other than the peculiar cases of Poisson distribution or combined Poisson distributions, the number of events in a given time is obtained from the knowledge of the law of the renewal time and not the reverse. Here we know first the distribution of the number of events and we are obliged to add some hypothesis to come with an approach of the renewal time distribution:

We suppose that w(t) is a random process with a correlation radius in time much larger than 1/F. With this assumption, F will last for a sufficient length in time to behave as a real firing rate for which the relation F=1/T does apply. The period T is the inter arrival time. The probability density function of T, calculated from the pdf of F, is:

$$g(T) = \frac{1}{\sigma T^2 F_D \sqrt{2\pi}} \exp\left[-\frac{\left(1 - TF_D \, \partial U\right)^2}{2\left(T\sigma F_D\right)^2} \right]$$

Results

In Table **1.16**, we compare the computed proportion of T included within two values of time (in ms) with the experimental results:

Table 1.16. The random firing of a cat ganglion cell.

T(ms)	> 2000	1000/2000	500/1000	250/500	100/250	50/100	25/50	0/25
Computed	0.017	0.026	0.032	0.087	0.327	0.540	0.002	0
Measured	0.018	0.020	0.032	0.112	0.345	0.290	0.023	0

Appendix M: αEEG Waves

Surface potentials derived from the cerebral cortex can be thought of as engendered by polarized leaflets, of which the unit components are parallel dipoles formed by the long pyramidal neurons. This geometry leads to an observable addition of the individual potentials when the neurons are synchronized.

The very regular α waves are probably due to a regulating modulation maintaining the control potential of swift neurons at the value $U_{Thr} + 0.15$ mV. Then, most of the neurons are synchronized by mutual influence as long as they are not disturbed by individual excitations. As seen in Section 3.2, the mean firing rate of the regulated neurons is about 10 s^{-1}. The oscillations of the controlling feedback are of the order of the slow synapses constant of time, $T_A \approx 1$ s. This number are consistent with the α wave characteristics.

Other waves, probably due to thalamo-cortical interactions, are less easy to interpret.

REFERENCES

[1]   Cajal RY. Histologie du système nerveux de l'homme et des vertébrés. Généralités, moëlle, ganglions rachidiens, bulbe et protubérances. In: A. Maloine, Ed. Paris, 1909-19112; pp. 986-993.

[2]   Skou JC. The influence of some cations on an adenosine triphosphatase from peripheral nerves. Biochem Biophys Acta. 1957; 23(2): 394-401.

[3]   Hodgkin AL, Huxley AF. A quantitative description of membrane current and its application to conduction and excitation in nerve. J Physiol London 1952; 117(4): 449-544.

[4]   Correa AM, Bezanilla F. Gating of the squid sodium channel at positive potential: single channels reveal two open states. Biophy J 1994; 66(6): 1864-1878.

[5]   Hamill OP, Marty A, Neher E, Sakman B, Sigworth FJ. Improved patch-clamp techniques for high-resolution current recording from cells and cell-free membrane patches. Pflügers Arch 1981; 391(2): 85-100.

[6]   Changeux JP, Podleski TR. On the excitability and the electroplax membrane. Proc Natl Acad Sci USA 1968; 59(3): 944-50.

[7]   Takeuchi A, Takeuchi N. A study of theaction of picrotoxin on the inhibitory neuromuscular junction of the crayfish. J Physiol 1969; 205(2): 377-91.

[8]   Chandler WK, Meves H. Voltage clamp experiments on internally perfused giant axons. J Physiol 1965; 180(4): 788-820.

[9]   Eccles JC. The Physiology of Synapses. Berlin. Springer-Verlag ed, 1964; p. 316.

[10] Llinas R, Sugimori M. Electrophysiological properties of *in vitro* Purkinje cell dendrites in mammalian cerebellar slices. J Physiol 1980; 305: 197-213.

[11] Gasser MJ, Erlanger J. The nature of conduction of an impulse in the relatively refractory period. Am J Physiol 1925; 73: 613-35.

[12] Eckert R, Tillotson D. Potassium activation associated with intraneuronal free calcium. Science 1978; 200(4340): 437-9.

[13] Krylov BV, Makovsky VS. Spike frequency adaptation in amphibian sensory bibres is probably due to slow K channels. Nature 1978; 275(5680): 549-51.

[14] Sehgal A. (Ed), Molecular Biology of Circadian Rhythms, Hoboken, NJ: Wiley-Liss, 2004; p. 296.

[15] Hartline DK, Cooke IM. Postsynaptic membrane response predicted from presynaptic imput pattern in lobster cardiac ganglion. Science 1969; 164(3883) 1080-5.

[16] Fessard A. Propriétés rythmiques de la matiètre vivante. Paris. Hermann ed. 1936; p. 65.

[17] Hugelin A, Bertrand F. The pneumotaxic system. Arch Ital Biol 1973; 111(3-4): 527-45. Review. French.

[18] Panda S, Hogenesch JB. It's all in the timing: many clocks, many outputs. J Biol Rythms. 2004; 19(5): 374-87. Review.

[19] Wright RH. Why mosquito repellents repel. Sci Am. 1975; 233(1): 104-11.

[20] Barlow HB, Levick WR. Changes in the maintained discharge with adaptation level in the cat retina. J Physiol 1969; 202 (3): 699-718.

[21] Hubel DH, Wiesel TN. Ferrier lecture. Functional architecture of macaque monkey visual cortex. Proc R Soc London B Biol Sci 1977; 198(1130): 1-59.

CHAPTER 2

Men, Animals, Machines

Abstract: From the simplest animal to the man, the structure of neurons does not strongly change. But the number of neurons increases continually: sea anemone owns 1 neuron, frog and honeybee 10^7, man 10^{11} with a three layered brain (limbic system, thalamus, cortex). Although human behaviors are very sophisticated, frog behaviors can be described by simple software. Thus, we had to compare brain with computer. Neurons are 10^6 times slower than electronic chips. Brain capacity for data storage is very weak when compared with the newest machines. Thus, books and magnetic memories appear as extensions of human memory. But the massively parallel circuitry in brain, the specialization of cortical areas and the modulating interactions between the three layers allow to animals and men a global efficiency not achieved by machines. Artificial intelligence (which mimics the human one) allowed for instance the building of almost autonomous automatic pilots. But a man has to intervene to choose the destination. Self learning machines have been conceived for instance to read postal codes. But men have to furnish them samples of each figure: robots can do anything except to choose the goal of their working.

Keywords: Artificial intelligence, behaviors, computers, control, cortex, data storage, evolution, frog, honeybee, individual preservation, large systems, limbic system, locomotion, memories, number of neurons, sea anemone, social instincts, spinal cord, thalamus, Turing machines.

2.1. Brief Description of the Human Nervous System

Main facts: General scheme, size of the cortical sub-units.

The human nervous system is made of 90 billion neurons (20 billion in the neocortex, 40 billion in the cerebellum) and probably 10^{15} synapses.

In an over simplified description, the human brain is made of a spinal cord and of three strata, the one with the greatest number of neurons is outside (Fig. **2.1**).

Between the spinal levels and the brain, we find an intermediate level, sometimes called "reptilian brain". It is mainly made of some motor nuclei, of the brain stem and is prolonged by the spinal cord. It directly controls the vital functions: cardiac frequency, lungs frequency, arterial tension, vigilance and sleep.

The first brain stratum is sometimes called "paleo-mammalian brain". It is made of the hypothalamus, the septum, the amygdalae and the hippocampus. It directly controls the emotions, the neuroendocrine interactions, the main behaviors (nourishment, defense, sexual activity) and plays an important part in memorizing process. Stimulations of well-chosen places in this stratum induce sham rage or hyper sexuality in cats or transform a furious bull into a pacific animal.

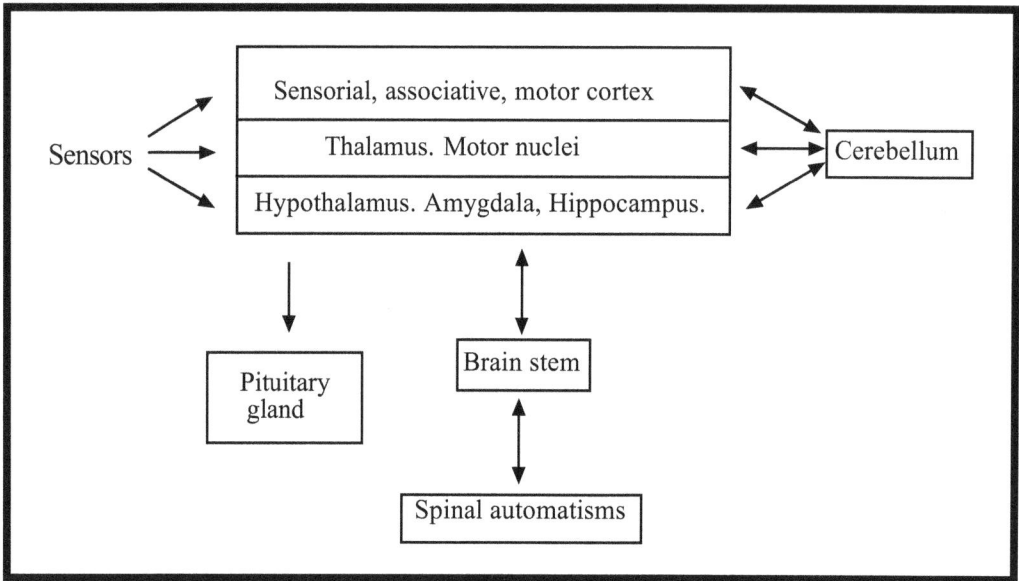

Fig. (2.1). Rough scheme of the human nervous system.

The second stratum is made of the grey motor nuclei and of the poorly known thalamus (corpus striatum, pulvinar, palladium, sub-thalamic nuclei, nucleus niger).

The third stratum is the neocortex, in charge of the cognitive functions.

In the poorly understood Alzheimer disease, some metabolic protein unfolding abnormalities cause a loss of cell membrane integrity, then its death. The damages reach the hippocampus first, then the associative cortex.

2.1.1. Each Stratum Modulates the Upper Ones

The hippocampus generates the exciting modulations required by the cortex to memorize. As shown by Korsakoff in 1889, failure of the hippocampus causes the impossibility to remember anything new.

The thalamic nuclei are in charge of the threshold modulation of the cortex. The corpus striatum receives signals from the associative cortex and from the nucleus niger. The nucleus niger receives signals from the premotor cortex and from the corpus striatum. The corpus striatum acts on the other basic nuclei, which act on the thalamus. When the nucleus niger is impaired (Parkinson disease), the thalamus is in excited state and cortical neurons at rest are maintained at a control potential definitely lower than the threshold. Specific excitations have to be greater than usual to trigger the firing of a neuron. Thus, motions are rarer, lower and with a less efficient regulation (shaking appears) as usual. When the corpus striatum is impaired (Huntington disease), the thalamus is overexcited; the control potential at rest of the cortical neurons is too high; some competitive inhibitions can be ineffective and unwanted motions appear.

2.1.2. Each Stratum Sends Signals to the Lower Ones

The pulvinar sends signals to the amygdalae (which governs rough behaviors such as fear). The frontal cortex sends signals to the deeper stratum (which induces observable phenomena such as sweating).

2.1.3. Parallel Processing

Each stratum carries an analysis of the sensorial signals. Thus, a rough visual analysis made by the pulvinar goes with the detailed analysis made by the cortex (see Section 3.1).

2.1.4. Size of the Various Sub-Units

The deep nuclei (pneumotaxic system, circadian master clock) have simple functions. A redundant circuitry confers a great reliability on them. They are usually made of 10^4 or 10^5 neurons.

The cortex is an assembly of sub-units, each of them made of about 10^6 elementary components (the columns), each column containing about 10 neurons. As the cortex is made of 10^{11} neurons, it contains about 10^4 sub-units, each of them in charge of an elementary operation. The cortex is usually divided into 40 Broadman's areas; each of the area is a set of about 250 sub-units. (All these numbers are only rough orders of magnitude).

2.1.5. Conservation of Vicinity

There is a strong topological relationship between the areas of the body and concerned areas of the brain. Neurons that deal with a specific part of the body are neighbors; and if two clusters of neurons are related to parts of the body that are close together, one cluster is not far from the other. (These properties are the bases of a theoretical device, the Kohonen self-organizing maps (Appendix G of Section 2.3).

2.2. Animals and Human Beings

Main facts: Relation between the increase of the number of neurons and the sophistication of the observed behaviors.

2.2.1. Anatomical Aspect

Some sea anemones count only one neuron that simultaneously acts as a sensor and as a motoneuron. When directly excited, it orders tentacles closure.

The frequently studied leech owns only 10^4 neurons. Honeybee owns 10^6 neurons, frog 2.10^7 of them. The most primitive mammal (the tanrec, a kind of hedgehog with a long nose which lives in Madagascar) enjoys the ownership of 5.10^8 neurons, the cat 10^9, and the gorilla 2.10^{10}. The nervous system of man is built from around some 9.10^{10} neurons (200 times more than the tanrec).

That evolution comes with a progressive increase of the number of the brain strata, then, in mammals, by a general increase of the ratio of the number of neurons in an animal to the number of neurons in the same area of the tanrec brain. Note that this ratio is not homogeneously distributed among the cerebral areas [1]. The main part of this increase is devoted to the neocortex (with 200 times more neurons in human neocortex than in the tanrec one). We observe also a noteworthy increase of the cerebellum, the diencephalon and the striate nuclei: a factor 9 for gorillas and 22 for men. This increase is associated with standing up and to the progress of motor ability, especially of the manual skill.

We would have now to look at the relation between these numbers and the evolution of the various behaviors. But we confess that we have never seen a tanrec alive and that we are not familiar with gorillas. Ultimately it will be more realistic (wiser) to compare man with the well known dogs and cats.

2.2.2. From the Sea Anemone to the Frog

2.2.2.1. The Sexual Life of the Sea Anemone

The one-neuron system of the sea anemone is not very efficient. This primitive animal owns (as plants) a sexual system controlled by hormones allowing accumulating and ejecting seeds. In place of a sexual instinct, it is driven by a simple automatism. In more advanced species, intromission leads to a huge economy of seeds, but requires a nervous device to recognize a female of the same species, a sexual instinct to control the motions and some social instincts to solve the conflicts.

2.2.2.2. Two Main Bifurcations in the Evolution of the Nervous System

2.2.2.2.1. The Outcome of Locomotion

Mobility requires nervous devices assuming the control of posture and motion, the (random or optimized) choice of a direction and often a navigation device using visual or fragrant beacons, sun direction and flight duration in bees, magnetic sensors in pigeons, mental mapping in dogs or men.

2.2.2.2.2. The Strategy of Individual Preservation

The nervous systems of individuals of the same species are often so strongly linked by pheromones that the higher hierarchical level is not the animal but the society.

Social insects (bees, ants) behave as if they own a common brain. The society is the real entity. The queen assumes the reproduction; the workers are only almost autonomous units without sexual instinct. The goal of the optimization process is to preserve the society at the detriment of individuals. While a queen lives several years, a summer worker lives only one month. It is first a nurse, then a builder, at last a supplier and dies. Pheromones control these transformations. The fast turnover, a real slaughter, allows to constantly adjusting the population to the needs of the bee-hive.

With or without pheromones, vertebrates own social instincts. The social behaviors of wolves or baboons are well known. But the social links help to preserve individuals. The entity is the animal: each of them owns a sexual instinct and the goal of the optimization process is to preserve any of them. Fear, mutual aggressiveness and general behavior responding to pain, all of them unobserved in insects, are associated with individual preservation.

2.2.3. From Frog to Human Being

2.2.3.1. How Does a Frog Work?

The frog, a primitive vertebrate, recognizes three classes of objects, worm type, fly type and predator type such as vultures. Electrophysiology shows that its retina contains neurons specialized for vultures (or other big animals as dogs), flies and worms. The sight of a fly triggers a stereotyped behavior: the frog turns its head, catches, swallows and wipes its mouth. This sequence is carried out till its end even if the prey is missed. The sight of a vulture leads to a choice between staying motionless (and almost invisible) or jumping towards a shelter. It would be easy for an engineer to build with standard neurons a device working as a frog.

2.2.3.1.1. Discussion

A fly is recognized by its size and its velocity. By summing and subtracting the signals of photoreceptors, the neurons of the third stage of the retina have a receptive field of which the diameter is adapted to the fly size. It is a transient swift neuron, the time constant of which is adapted to the fly velocity. Thus, a "fly neuron" emits a burst of spikes at the sight of a fly, but not at the sight of a static object or of a vulture.

The excitation of any fly neuron triggers the eating behavior. The excitation of any vulture neuron triggers the fear behavior. The solution of conflicts is simple: fear behavior has an absolute priority, it inhibits the eating behavior.

The behavior order governs a sequence of elementary acts, each of them controlled by a feedback device and modulated by the eye in direction of the prey. In the sequence, the end of each elementary act triggers the beginning of the following one. Note that this analysis requires a double visual pathway, both using operations OR: the first (a fly at the right OR a fly at the left) triggers the behavior; the other (fly OR vulture at the left) controls the direction of the motions.

2.2.3.1.2. From Frog to Human Being

The same general scheme (identification of an object, choice between incompatible behaviors, execution of a behavior) applies to all the vertebrates. But, from frog to man we observe a continuous sophistication of the processes.

a)   Elaborate coding and accurate analysis with the benefit of progressive learning increase the discrimination ability. While a male lizard reacts to the smell of a female without identifying her, the perch, a moderately advanced fish, roughly recognizes and prefers its usual partner. Man seems able to recognize any object (when the frog recognizes only three classes of them).

b)   Generalized instincts lead to a constantly adapted choice of behaviors.

c)   Behaviors are fragmented into a great number of partial behaviors, each of them owning its associated drive signal and its almost autonomous control system: a man does not wipe its mouth if he has not eaten. This fragmentation leads to the appearance of gradations in the main behaviors: while some fishes know only two conducts, to flee away or to fight to kill, more advanced animals are able to discriminate between a threat (a potential attack) and an actual attack.

2.3. Machines and Men

Main Results: Although neurons are very poorly efficient chips, their specialized wiring allows results that the multipurpose computers are not able to reach. Brain is not similar to a computer, but to the control device of a large industrial plant. In the brain, the number of neurons devoted to a given task is close to the theoretical minimum.

Introduction: Both brain and computers transmit, transform and store data. Scientists and engineers have developed control theory, information theory [2], computer science [3], filtering, and tracking and classification algorithms. They have been able to create artifacts that in some extent behave like human beings or animals, the most popular and the most phantasmal of these accomplishments being the robots. The buildup of such a corpus of knowledge should be of great help to understand and describe how the brain is actually working. Some parts of the nervous system are strikingly similar to a computer or to an artificial neurons set. But the whole brain is strikingly different of both. A meticulous discussion is needed.

2.3.1. Is the Brain a Computer?

In a first step, we compare some quantitative characteristics of the human brain to those of common PC's and of the most powerful 2013 supercomputer, the Tianhe-2 (Milky Way-2) developed for the Chinese University of Defense Technology by the national super computer center in Guangzhou. It is impossible to figure out the performance of a computer by referring to only one characteristic. All together performance depends of several factors: clock frequency, number of cores, architecture and software organization…

2.3.1.1. Clock Frequencies

If we compare the clock frequency of an average PC, 3.10^9 Hz, to the neuron working frequencies, from less than 10^2 to 10^3 Hz, we get the sense that the brain is highly handicapped by the slowness of its components. PC's are awfully better than brain to do numerical calculations. It takes to the brain some seconds to multiply a number of two digits by another two digits number but a computer takes few instructions to achieve the same computation (it takes less than 10^{-8} seconds in a PC). To evaluate and compare the computing powers of their machines, engineers test their ability to compute the solution of a standard linear problem (the LINPAC bench mark). The results are specified in Floating point operations per second (Flops) which is a meaningful and reliable unit of measurement only for numerical computing.

A PC may achieve several 10^8 Flops; 4.10^{10} Flops had been measured in Tianhe-2. In the brain (which adds two numbers with several decimals in more than 1 second), the computing power would be less than 1 Flops. Obviously, the comparison has no meaning.

2.3.1.2. The Computing Power of the Brain

We can easily compute the number of nervous spikes per second: competitive inhibitions let fire simultaneously only 10^{-3} to 10^{-2} of our 2.10^{11} neurons. Their mean firing rate is about 300 s^{-1}. Thus, we expect between $0.6\ 10^{11}$ and $6\ 10^{11}$ spikes per second. But we do not know how many spikes are needed to transmit meaningful data. Several authors have proposed different approaches to estimate the power of the brain in terms of instructions executed in one second, a quite floppy and unreliable unit of measurement (Mips: Mega instructions per second). These approaches are based on general physical considerations and the natures of the operations of the brain that are involved in the estimation process are not clearly identified. The resulting guesstimates fluctuate from 10^7 Mega instructions per second to 10^{13} Mips. To be compared with the results of super computers bench marking, these values should be convert into compatible units, a not easy task and probably not worth to undertake. Moreover, the extent of the interval of fluctuation of the estimates of the brain power betrays our lack of knowledge of its performance. This lack of knowledge precludes the possibility to build any sensible quantitative relationship between the computing powers of both computer and brain.

2.3.1.2.1. Data flow

Comparing the data flow going into the brain with the data flow going out a super computer is perhaps significant: About 10^9 Bytes per hour in a DVD; Production of Blue Gene, a former top 500 super computer, 4.10^{13} Bytes/hour; Human visual inputs (2 eyes, 10^6 optic fibres from each eye, 30 signals per second): 10^9 octets/hour.

2.3.1.3. Long-Term Memory

Assuming that less than 10% of brain synapses are conditional ones, the capacity of brain memory would be less than 10^{14} bits (1.25 10^{13} Bytes) while the memory of the hard disk of a PC is more than 10^{11} Bytes; of Blue Gene/L: 10^{17}Bytes; of Tianhe-2: 10^{17}Bytes. The human memorizing capacity is not very big.

Now, two different strategies appear: computers gather and store data while the human looks at the meaning and the interest of the signals in order to avoid storing useless data. Look for instance at a medical X ray scanner. The X ray source supplies a set of brute picture for each patient. The computer simultaneously uses all of them to evaluate the absorption coefficient in a huge set of (x, y, z) localizations and stores the data of all the patients. The physician is unable to read such a large amount of results. Auxiliary software uses the data to build synthesized images of various sections of the patient body. The physician analyses them and reduce them to a small set of words (to formulate a diagnostic) and forget quickly the diagnostics of most of his patients. In fact the physician filters most of the incoming information. We shall see later on that this filtering process is due to an emotional center, the limbic system. This process let the brain to work efficiently with much less memory than super computer.

A main difference in the way computer and brain manage data through memory is the following: Computer does store and does retrieve and delete information almost instantaneously. As we will see later in the text, brain registration is delayed, reading is almost instantaneous and deleting does not exist. A progressive and slow forgetting mechanism replaces the deleting process.

We conclude that computers can be used as an artificial fourth stratum of the brain: they marvelously transmit, transform, store and mine data, but they do not replace the natural strata.

2.3.1.4. The Architecture

As a first difference, the brain is a more highly parallel device than PC or super computers:

2.3.1.4.1. Number of Sub-Units

1 core-processor in PC; 3.12 10^6 cores in Tianhe-2; about 10^4 sub-units (each of them made of about 10^6 columns) in the human cortex.

But the main difference bears on the specialization: PC and Tianhe-2 are multipurpose machines; they accept any software. On the opposite, each neuronal sub-unit of the brain (for instance a visual area) is built to carry out a specific operation.

2.3.1.4.2. Von Neumann Computers

During World War II, J. von Neumann designed the first computer. All the modern ones derive from it. As reminded in Appendix A, the architecture of a conventional von Neumann computer is based on the association of a finite automat, a processor unit, and a directly addressable memory. Any piece of information, either data or instruction, is stored in the memory in a specific location clearly identified by an encoded address and is retrieved by decoding its address. Data are coded information related to the initial conditions of the problem to be solved, to the intermediate results and to the final results. Instructions are encoded orders to control the computing process.

The basic computing process performs the following cycle of events

1.   Fetch an instruction from memory

2.   Fetch any data required by the instruction from memory

3.   Execute the instruction (process the data)

4.   Store the result in memory

5.   Return to step 1

On the opposite, each area of the brain is devoted once for all to a precise instruction. Such architecture favors parallel processing. But it forbids changes of software.

The problems that are possible to solve by using von Neumann computers are those which can be formalized in term of an algorithm. An algorithm is an accurately defined step by step procedure, which will guarantee the answer. It can be broken down into intermediate statements and finally may be reduced to a list of executable instructions. Problems that can be solved in such a way are, for example, numerical solution of differential equations or search for an item in a file.

Moreover it is possible to define algorithms, which are able to process strings of symbols according to the rules of some formal system. It was the hope of the proponents of the symbolic Artificial Intelligence that all knowledge could be formalized in such a manner and that any intelligent process could be reduced to the manipulation of symbols according to rules and that this manipulation could be implemented on a von Neumann machine (conventional computer). The success of the symbolic approach depends directly on the faculty to find an algorithm to describe the solution to the problem. But it is not always possible to reach in advance the level of accuracy and great detail that would be necessary to define a workable algorithm (the base of rules), and above all, it is not always possible to identify and to take in account and to depict with the required accuracy the total amount of knowledge (the base of facts and the base of knowledge) that would be necessary to reach a satisfactory solution.

It turns out that many everyday tasks we take for granted are difficult to formalize in this way. For example, take our visual (or aural) recognition of things in the world; how do we recognize handwritten characters, the particular instances of which, we may never have seen before, or someone's face from an angle we have never encountered? How do we recall whole visual scenes on given some obscure verbal cue? The techniques used in conventional computer are too constrained or to cumbersome to account for the wide diversity of associations we can make.

Therefore, even if in some restricted and well defined area, the conventional computer may be a good tool to generate some adaptive behaviors that are quite similar or even more efficient than the behavior generated by a human being, the conventional computer can be used neither as a model of the brain neither for a complete description of its working process.

2.3.1.4.3. Number of components

Number of semi-conducting junctions in a PC > 10^9; in Blue Gene > 10^{14}; number of human neurons: 2.10^{11}

2.3.2. Artificial Neural Networks

An alternative approach to Artificial Intelligence, using artificial neural networks, has been proposed from the beginning of the time, when researchers started to pay attention to the possibility to create automats which would behave in a more or less intelligent way. These theoretical works throw out some light on the results of physiological observation.

A neural network is an interconnected set of very simple processing cells (nodes) which are supposed to roughly mimic the performance of an animal neuron. The style of processing performed by neural network is completely different of the symbol processing performed by Von Neumann computers. It is more akin to signal processing than digital computation. An output is produced by combining a large amount of signals and not by executing sequences of instructions stored in a memory. Information is stored all over the net in the relative power (weights) of the arcs interconnecting the nodes rather than in predetermined locations in a specialized device. The weights are supposed to adapt themselves to produce the right answer when the neural net is presented a configuration close to a member of a family of training configurations. The result of a computation appears as a pattern of activity across a number of nodes and not as the content of specific locations in a computer memory.

Neural net are not pure mathematical abstractions. Some have either been implemented on VLSI as electronic circuitry or have been emulated on universal computers to be used in many areas to solve practical problems: from control system to video game passing through market forecast, signal processing, pattern recognition, classification or clustering of objects and so on. Many approaches have been proposed or followed within the discipline of neural nets and researchers are still opening new roads towards networks with better performance characteristics. Appendix B discusses the subject in some more details.

Neural nets, by their connectionist nature, are copying much more closely the massively parallel architecture of the brain than the von Neumann computer. The architecture of an Artificial Neuronal Net (ANN) is the pattern of interconnections among the artificial neurons in the same way as the architecture of the brain is the pattern of interconnection among biological neurons. But basic differences with the brain appear as soon as we consider the key building block of an artificial neural net, the artificial neuron.

The oldest model of neuron (the Mc Culloch and Pitts neuron [4]) mimics the behavior of all or none physiological neurons connected by all or none swift synapses. Each synapse linking the neuron i to the neuron j of the next layer is characterized by its synaptic weight W_{ij}. A neuron fires if the weighted sum of all the inputs on its exciting synapses is greater than a threshold (Appendix C, D): The transfer function of the input to the output is a harsh step. A variant is used in multi layers networks where a differentiable transfer function is necessary to make possible the application of a training algorithm (back propagation of the gradient). In these cases, the transfer function takes the smooth form of a sigmoid or of a hyperbolic tangent. This is, up to now, the only variant introduced in the model of Mc Culloch and Pitts.

It should be noticed that the kind of neuronal operation, here above described, only matches the operation of aperiodic neurons linked by swift synapses. All along this book until now, it has been shown how many different types of neurons and how many basic functional circuits are building blocks of the brain. Moreover, some areas of the cortex are more or less specialized in performing certain functions. Therefore, if the brain is a connectionist device, it does not interconnect standardized aperiodic neurons to form a somewhat homogeneous mesh similar to an artificial neural network. The neural nets may sometimes give a better image (see Kohonen maps in Appendix G) of the behavior of the brain than von Neumann computer but, for sure, the brain is far more complex than the most complex artificial neural net.

2.3.2.1. The Hopfield's Limit

Using a model of mathematical neurons with smooth transfer function (Appendix E), J.J. Hopfield [5], evaluated an optimistic theoretical limit of the storage capacity of large network including at least several hundreds of vertexes for an error rate of 1%. (This limit is for instance, for a recognizing picture device, the number maximum P of pixels which can be recognized). Let N be the number of neurons; the limit is P = 0.14 N. If we compare for instance (Section 5.1) the number of neurons in the olfactive cortex and a rough evaluation of the number of recognized smells, we find an order of magnitude compatible with the theoretical limit. Thus we are led to this important conclusion: the brain circuitry is optimized to minimize the number of neurons required by the data processing.

2.3.3. Brain and the Control of Large Systems

2.3.3.1. Layered Architecture for the Control of Large System

The nervous system is more than a pure multipurpose computing machine. The brains of the most evolved animals are large systems dedicated to control the complicated processes of animal behaviors. As such, nervous system (including the brain) retains some architectural characteristics of control systems that take care of large plants like electricity transport and distribution networks, oil-plants, networks of self-controlled automatic subways, air plane supervision and control (Appendix H).

In order to tackle the problem of controlling large and complex systems, engineers often intend to break up the global process into sub processes that can be control locally (decomposition). The solutions obtained at the local level must be more or less coordinated in order that their juxtaposition adds up to the solution of the global problem (coordination and decentralization). Of course this approach does make sense only if the sub problems are easier to solve that the main problem. It may happen that the sub problems need also to be broken up leading to several stages of decomposition and to a layered architecture of the control system. One of the advantages of such an approach is that the increase in computing power implied by an increase in size of the main problem is much smaller than the increase that would be required by a fully centralized control. In many cases the subsystems already exist as physical entities and the decomposition process is fully constrained. If the decomposition is not fully constrained, the question of the best split of the system arises.

When the control is added to an existing physical structure, the advantages of maintaining decisions close to the physical components are:

-       To reduce the flow of information that is necessary for the performance of the global system

- To improve the behavior of the global system in case of local failure (fail safe behavior)

- To make possible an evolving approach for the implementation of the control mechanism. (A decentralized structure authorize to add new components to a plant without questioning the whole control system or to improve the overall control without deeply modifying the local mechanisms)

It seems that nature has been well aware of these advantages when it went to develop the central nervous system of animals. The first advantage is relevant to the principle of economy of neurons, the second to survivability, and the third to the process by which the evolution increased the complexity of nervous systems in animals.

The most well known mechanisms of coordination are the coordination by resources and the coordination by prices which are at the foundation of classical economic models.

2.3.3.1.1. *Example of an oil-plant*

For example an oil-plant is made of several almost autonomous units (tanks, processing towers...). The scheme of the controlling system has to respect this fact. A pragmatic approach led to build multilevel control devices. Mathematical studies came then to justify these ideas.

Each almost autonomous unit is associated with a local controlling device which is in charge of the security and of the adjustment of the regulated parameters to the commands. The set of local devices is the lower hierarchical level. Higher coordinating levels receive information from the lower level (for instance the amount of products in stocks) and from the outside world (for instance the forecast of the market demand) and computes the orders transmitted to the lower level.

2.3.3.2. Lower Level Devices in the Nervous System

Can we detect such a hierarchy in the nervous system? Some parts of the body (such as heart or lungs) are clearly almost autonomously controlled units. Another example is given by skeletal muscles: they are embedded in a low level spinal hierarchy controlling the coordination between agonist and antagonist muscles or the automatic walking motions of a new born baby.

In primitive animals, such devices are driven by an autonomous circuitry with only an ON/OFF order coming from a higher level (this is the case of automatic walking, the standing up position creating an ON order). A progress caused by evolution leads to insert weak exciting or inhibiting modulations in several places of the circuitry. Thus, the heart is autonomous but its frequency is modulated by orders coming from higher levels. In the next step due to evolution, strong modulations are inserted in the primitive circuitry. In grown up peoples, automatic walking is fully inhibited and walking, running or dancing are controlled by higher levels orders transmitted through the primitive autonomous circuitry: the hierarchical organization is somewhat masked.

APPENDICES TO SECTION 2.3

A: Turing Machines

B: The artificial intelligence

C: Artificial neurons

D: Artificial neurons nets

E: The Hopfield devices

F: Boltzmann machines

G: Kohonen self-organizing maps

H: Adaptative Resonance Theory of Neural Networks

Appendix A: The Turing Machine

To understand the roots of the difference between our universal computer and the brain, we have to come back to some basic attributes of what we usually call a computing machine. The theoretical foundation of digital computing is the conceptual Turing Machine (TM) which has been demonstrated to be equivalent to a Direct Access Machine (DMA). DMA is the abstract model of the computers that are designed according to the Von Neumann architecture. Therefore results on the capabilities and limitations of our universal digital computers, all of the Von Neumann type, derive from analyses carried out on the Turing Machine.

Without entering in a full and rigorous description of the TM we will just insist on the main characteristics of its operation and on its capabilities. The simplest Turing machine is in essence a finite automat augmented with a memory; the memory is a tape of infinite length on the right and that is ruled in squares each of which can carry a single symbol (think of a magnetic tape). The TM works on a finite set (an alphabet) of symbols, one of which is a "Blank" symbol. Any finite collection of symbols of the alphabet is a word. A part of the set of all possible words is a language. A TM is programmed through three functions of its states and of its alphabet: a writing function, a transition function which changes the state of the machine, a shift function which moves a reading/writing head one step left or right. A computing stage is made of the following cycle of operations:

-       *The head read the symbol a included in the square in front of it.*

-       *The head write in the same square the symbol b which is determined by the printing function from the previous symbol a and the state q of the machine.*

-       *The shift function move the head one step left or right pending on q and a.*

-       *The transition function change the state q of the machine to a new state q_1 according to the present state q and the symbol a.*

When the head reaches one of some specific states, the terminal states, the head does not move anymore and the result of the computation can be read. The machine may also never halt.

Although it appears to be a very crude architecture, certainly much simpler than the architecture of human brain, the Turing machine is able to carry out very complex tasks if only sequential manipulations of symbols are involved. In fact any operation on words of symbols of the alphabet that can be achieved by a finite mechanical sequence can be performed by a Turing machine. These operations make up the class of computable functions. *If Γ is a part of the TM alphabet that does not include a "Blank" symbol, and Γ^* the language build on Γ, a computable function F is a function that is partially defined on a subset of the TM words and takes its values in Γ^*. If the machine halts after a finite sequence of mechanical actions (instructions), F is said to be computed by TM. Its value is obtained as the first word without blank that can be easily identified at the left end* *of the final configuration. The relevant sequence of actions is called an algorithm. If the machine does not halt, F is not defined for the given inputs.*

The architecture of the Von Neumann computers that we use on a day to day basis is also an association of an automat (processor) and of a memory. As already mentioned, this kind of computer has been proven to be equivalent to a Turing Machine. Therefore it benefits from the same capabilities and suffers the same limitations than the Turing Machine. It can only deal with computable functions.

Appendix B: The Artificial Intelligence

But our brain does not only work by manipulating symbols and with computable functions. This is made obvious by the quite poor results obtained in some areas when trying to copy intelligent behaviors through the symbolic computation of expert systems. *These systems can be successful only if the following conditions are fulfilled:*

- *The decisions, which the system may have to take, must only depend on a restricted and well defined set of variables.*

- *The values, which the variables can take, should be known and should be expressed with words.*

- *If some form of uncertainty is accounted for, it is modeled by functions of the recognized variables alone.*

- *The manner, in which the decisions depend on the values of the variables, must be known and must be computable.*

To solve a problem within the context of the symbolic approach it is necessary to supply an algorithm with the complete knowledge of the domain of interest. Consequently expert systems are confined to suitably restricted areas for which it is possible to collect and to formalize the whole set of pertinent facts. This restriction is at the origin of some frustration about the so called Artificial Intelligence and about its inability to cope with trivial problems, which are easily mastered by human brains. Brain is by a far much better than symbolic AI at associating or clustering a mass of noisy or ill defined data to find out possible solutions and to select one which is acceptable. Anyway the frustration about AI comes more from the excessive optimism of its first promoters than from a complete lack of success.

Appendix C: Artificial Neurons

An artificial neuron is defined by: a vector of inputs $x_{i=1,2,...,n}$, a set of weights on each input $w_{i=1,2,...,n}$, a threshold θ, a transfer function f, an output signal s.

The "activation" of the neuron is the weighted sum $a = S_i w_i x_i$ and the output s is a function of the activation: s = f(a).

A clock activates the system at times t, t+Δt, t+2Δt... The activation of the neuron at time t will decide of its output s at the next time.

The simplest transfer functions f(a), (the Mc Culloch and Pitts one) is: $s = 0$ if a < q and $s = 1$ if a > q. (Note that 0 may obviously be changed for -1 to obtain a symmetrical transfer function).

Other usual transfer functions are continuous: The ramp function: $s = f(a) = ka$, the hyperbolic tangent function Th(a) function and the sigmoid function

$$s(a) = \frac{1}{1 + \exp(-ca)}$$

Appendix D: Artificial Neurons Nets

The architecture of an Artificial Neuronal Net (ANN) mainly is the pattern of interconnections among the artificial neurons in the same way as the architecture of the brain is the pattern of interconnection among biological neurons.

Feed forward architectures are most often used to build ANN. This type of ANN is an example that fits in the class of pattern association devices. The pattern association paradigm is one in which the goal is to build up an association between patterns defined over one subset of units and other patterns defined on another subset of units. In the feed forward architecture the connections between neurons are generally unidirectional. The network is hierarchically organised along layers of nodes (neurons) and information is transferred from one layer to the next. The first layer receives the input and the last layer contains the result of the computation, the output. Between the input and the output layer, "hidden layers" process the information. The simplest feed forward networks are made of only one input layer and one output layer. In fact, the perceptrons, the first models of operative neuron network [6], which were able to learn, were put together according to this scheme. But they suffered serious limitations: they fail to split a space of data in more than two regions; the two classes of data have to be linearly separable and the solution finally found by the perceptron is only an acceptable solution. Anyway, the work of Widrow [7] on his ADAptive LINear Element (ADALINE) and its learning algorithm opened the door to the learning by back propagation and to the multi layers networks. The main advantage of this last type of nets is their capacity to complete non-linear separations and so to enjoy a better ability to carry out some complex classification of data. The research community steadily improves this capability but it is out of the scope of this short introduction to mention all the more or less successful developments that are taking place. One of the first and best-known achievements of feed forward neural nets has been to succeed in developing an automat, called NETtalk, able to correctly pronounce an English text. Surprisingly, during learning, the net has progressively improved his talking ability from incoherence to babbling then to a correct pronunciation just like a baby learning to talk.

The great interest of artificial neuron nets is their inherent learning ability. It has been assumed that learning of physiological neurons is due to a change of synaptic couplings [8]. Such a phenomenon is easily introduced in mathematical neurons, less easily in standard computers. *This does not mean that universal digital computers are never able to learn. They can be programmed to identify changes in the parameters included in the model of a process and to adapt to a new situation. This has long been used in process control and more recently in programming artificial players of simultaneous games to make them able to choose winning mixed strategies against human players, for instance SAGACE (Solution Algorithmique Génétique pour l'Anticipation des Comportements Evolutifs). It should be noticed that the learning module of SAGACE is based on a very powerful heuristic using repetitively a very simple algorithm: the genetic algorithm. But the learning ability of von Neumann computers is highly application dependant.*

The learning capability of the feed forward net is achieved by implementing a computation algorithm that adjust each connection weight in such a way that the error between the output of the network and the required answer tend toward zero. At each step of the learning algorithm, the error between the existing output and the desired output has to be known. Therefore, two sets of configuration have to be submitted to the net and compared an input configuration and the desired output. This type of learning is referred to as "supervised learning" by likeness of this process with the process of learning under the supervision of a teacher. Several kinds of learning algorithm have been proposed; most of them proceed from two basic approaches: back propagation of error and competitive learning. Competitive learning is of special interest since the most elaborate functions of the brain often follow a competitive execution scheme (choice box). An example of competitive learning will be discussed later when the model of Kohonen nets will be exposed.

Appendix E: The Hopfield Device

But not all of our knowledge comes from learning by comparison and error correction. Since we are born, we have learnt and memorized a tremendous amount of things without even realizing that we were doing so. J. J. Hopfield, in 1982, devised a network that could theoretically learn and store data without supervision and is able to find a correct knowledge out of the proposal of a partial knowledge. *The Hopfield approach is relevant to the auto associative paradigm in which an input pattern is associated with itself. The goal here is pattern completion. The working of the Hopfield device is based on the dynamic of physical systems. At each time during its evolution, a physical system is described by a state vector X and, if several stable states $X_1,..,X_i,..,X_n$ do exist, the system will evolve to reach one of them. In each stable state, the energy of the system is at a minimum. If the starting point is $X = X_i + \Delta$, a point in the states space which is close enough to X_i, the system will evolve until $X = X_i$. Xi is told to be an attractor point of the system. One can think of $X_1,..,X_i,..,X_n$ as the information stored in the system, $X_i + \Delta$ as the partial knowledge and $X = X_i$ as the total knowledge. Any physical system of which the trajectories in the states space lead to a substantial number of locally stable states may be seen as a memory that is addressable by its content. Such a memorizing capability will be usable if and only if any set of prescribed input states can be forced into becoming locally stable states for the system. Hopfield has been successful in designing such a system with a fully connected network of Mc Culloch and Pitts artificial neurons. The matrix of the weights W is symmetrical and its diagonal is null since reflexive connections from a neuron to itself do not exist. X is the state vector of the network, its components are x_{ij}, the states of each neuron. The components x_j can only take two values either 1 or -1 (the transfer function of the neurons is a symmetrical step function sign(x)). By definition, the energy of the system is $E = -\frac{1}{2} X^T W X$ (here the threshold is neglected). The learning algorithm is based on the reinforcement rule defined by Hebb $W_i = X_i X_i^T - I$ where W_i is the matrix of weights that are forced by the input vector X_i and I is the unit matrix (the diagonal terms of W_i must be null). The dimension of an input vector is the* *number of neurons of the net n. The energy of the network in function of a vector X and for the set of weights W_i is:*

$$E(X) = -\frac{1}{2}\left(X^T X_i X_i^T X - X^T X\right) = -\frac{1}{2}\left(\left\| X^T X_i \right\|^2\right) + \frac{n}{2} \; since$$

$$X^T X = x_i^2 + \ldots + x_n^2 = n$$

The minimum of E(X) is obtained for X=X_i and X_i is an attractor of the system. If we input a vector X not too different from X_i and let the network free to evolve, it will evolve towards X_i. The network, which has been trained for the set of weights W_i related to X_i, has stored the information about X_i in its weights and is able to recover this information out of an incomplete input. By addition several references X_i can be overlaid on the network to create the same number of attractors. But crosstalk between the references may degrade the performance if they are too much correlated and too numerous. The best performance is obtained with fully orthogonal references. Hopfield estimated to 0.14 N the limit of the storage capacity of large network including at least several hundreds of vertexes for an error rate of 1%, where N is the number of neurons (a result confirmed by a theoretical study [9]). If the number of references is increased beyond this value, a catastrophic degradation takes place; the attractors have no more relations with the references.

Hopfield networks seem to have interested biologists because of the analogies of their memorization capabilities and the capabilities of the brain. Like the brain they are able to recognize incomplete or noisy information; like the brain they are subject to false memorization and to forgetfulness. Robustness also is common to the brain and the Hopfield net; performance of both systems smoothly deteriorates when cells are destroyed. But the Hopfield net supports a very approximate model of the cognitive performance of the brain and its homogeneous design has not much in common with the highly structured architecture of the central nervous system. Only some subsets of the cortex, micro-columns composed of some thousand of fully connected neurons, could be compared to the fully connected Hopfield net.

Hopfield net have also succeeded in finding a solution in polynomial time to the NP-complete problem of the travelling salesman. Since the work of Hopkins and Tank (1984) it has been proved [10] that the finding of the solution by a network of the size of the problem cannot be 100% guaranteed and that any improvement would cost an exponential growth of the network. The trend of the classical Hopfield net to converge towards local minima makes them not suitable for solving optimization problems that presume the search for a global minimum.

Appendix F: Boltzmann Machines

To overcome the limitation of the Hopfield net in solving optimization problems, Ackley et al., [11] have proposed a Hopfield net endowed with "visible" neurons and hidden neurons. The visible neurons are the interface between the network and the environment that specifies to it the vectors to learn or asks it to complete a partial vector. The hidden neurons are where the network can build its own internal representation. The probability of change in the state of a neuron is derived from the law that governs the change of energy levels in statistical thermodynamic. This local probabilistic decision rule ensures that in thermal equilibrium the relative probability of two global states of the net is determined by their energy difference and follows a Boltzmann distribution. So this class of artificial neurons nets is labelled Boltzmann machine. Playing with the parameter T of those laws, a dummy temperature, allows conducting the whole process of learning like an annealing process (simulated annealing). The simulated annealing is the stochastic mechanism by which the system may not be trapped in a local minimum and may reach a global optimum state.

Even if it attracts a great theoretical interest, its algorithmic intricacy makes the usage of the Boltzmann machine quite delicate. The multiple steps of computation and the cost of computing the different probabilities, which are required to carry out the learning process, have limited the use of this machine in practical applications.

Appendix G: Kohonen Self-Organizing Maps

Within the brain, the processing of sensorial information and of motion control is not randomly spread. There is a strong topological relationship between the areas of the body and concerned areas of the brain. Neurons that deal with a specific part of the body are neighbours; and if two clusters of neurons are related to parts of the body that are close together, one cluster is not far from the other. The work on self-organizing cards by T. Kohonen [12] stems from recognizing this topological characteristic of the brain. The model of self-organizing cards belongs to the class of networks trained through competitive algorithms. We shall review two members of this class, the already mentioned Kohonen net and the network developed in accordance with the "Adaptive Resonance Theory".

Kohonen network only uses one layer of neurons in addition to the input layer. *(At the beginning of the work on this kind of net, the model was developed to satisfy biologists more than engineers. The basic unit was not the neuron but the cortical column of 200 neurons).* Those two layers are often arranged along two dimensions even if it is possible to build networks arranged on more than two dimensions by simulation on digital computers. The particularity of Kohonen nets is to account for a notion of vicinity between neurons. The basic idea behind the conception of Kohonen nets is to manage the connections in such a way that all inputs related to one type of characteristic is connected to only one output neuron and that inputs of near-by characteristic are connected to near-by neurons. The objective is to build a topological relation between the space of data and the space of characteristics. Kohonen networks make use of only one layer of neurons generally organized on two dimensions.

Kohonen nets work by using three mechanisms: competition, cooperation, and adaptation.

For each input only one neuron must be activated; there is competition between the neurons and the winner takes all (best matching unit). Given $X = (x_1, x_2, ..., x_m)$ an input vector of dimension m, to each neuron i is attached a vector of weights $W_i = (w_{i1}, w_{i2}, ..., w_{im})$ associated with each connection between i and the input. The w_{ij} may be seen as the coordinates of the neuron i in the m space of the inputs. The neuron that wins the competition is the one for which the distance to the input X is minimum whatever is the choice made to express a distance in the space of input:

For instance: $|w_{1i}-x_1|+|w_{2i}-x_2|+...+|w_{mi}-x_m|$ *or* $\sqrt{\sum_{j=1}^{j=m}\left(w_{ji}-x_j\right)^2}$. *It is also possible to look for the maximum of the scalar product $w_i^T X$ and to select the neuron which is fulfilling this condition.*

In the same manner as in the brain, the activation of a neuron tends to influence its neighbours either by cooperating with their excitation or by inhibiting their activation. A function $\Phi(i,k)$ defines the coupling between the selected neuron i and neuron k. Kohonen proposes a Mexican hat type of function that takes care of inhibition at some distance of neuron i; but a Gaussian function is commonly *used. Adaptation through learning is based on a rule derived from the Hebbian rule. The purpose of the training is to try to push the vectors of weights as close as possible to the input vector. The strength of the adjustment of the weights varies in function of the distance between the neuron under consideration and the winner. The function Φ(i,k) modulates the value of the adjustment.*

$$W_i(t+1) = W_i\big(t\big) + \eta \Phi(i,k)\big(X - W_i\big)$$

As usual η is the value of the learning step. In the above equation η and Φ(i,k) may depend on the time and η may be different for the winner and the other neurons. Φ(i,k) forces the neuron k to occupy a position in the vicinity of i, if k is closely connected to i.

The algorithm includes the following stages: The matrix of weights is initiated with values chosen at random; an input vector X is entered; the neuron i that received the strongest signal is selected then the learning rule is applied to all the neurons. If the network has not yet converged the algorithm returns to the second stage. If and when the network has converged another input is proposed. It has been proven that for one-dimension networks, the convergence of the algorithm is guaranteed. For multidimensional network such a proof does not exist. If a two dimensions network is given inputs spread over a square, the network will converge to a configuration of neurons laid also on a square. Each neuron stands for a specific area and any input vector laid in this area will fire the related neuron and only this one. More generally, at the end of the learning process, the space of representation of the network corresponds to the space of the inputs. Areas that are close together in the input space are close together in the network space.

Kohonen networks are extensively used in many fields such as robotic (angular transformations), image compression, failure diagnosis *etc*. They are strong competitors for the classical multivariate analysis techniques. Their ability to put together pertinent representations such as semantic charts may be regarded as an embryonic capability to elaborate a form of concept.

Appendix H: Adaptive Resonance Theory of Neural Networks (ARTNN)

They have initially been developed to try to understand how an animal can learn to recognize the features of its environment and can learn new elements without forgetting those already acquired. Their main goal is to achieve dynamical classifications while trying to solve the contradiction between stability and plasticity. ART networks are hybrid nets acting all together as multilayer and self-organizing nets. Moreover the architecture of the network is evolutionary and adaptive.

In most neural networks two kinds of operations are clearly separated: During the operation of learning the weights are modified according to a learning rule; then, during the execution, the weights are frozen and are no more modified. If such a network is trained for a finite training set, and a new vector is presented as input that does not belong to the original training set, the training may be disturbed altogether, and training may have to be repeated all over. In an ART net the two operations are effected simultaneously. When learning, the net adapts itself to unknown inputs and creates new classes by adding neurons while saving almost all the information already stored.

The structure of the network makes use of one input layer that is also the output layer and of a hidden "comparison" layer. In addition the network includes long-term and short-term memories. The hidden layer works in accordance with the principle of the competitive activation ("winner takes all"), The selected neuron maximizes the scalar product between the input vector and its vector of weights. All the neurons of this layer are linked together by inhibiting connections. The neurons of the input layer have no connections between them and two specialized neurons play two different parts in the process. One of those controls the transition of the visible layer from the input to the output configuration; the other is a comparator that fires when necessary to inhibit the selected hidden neuron and to exclude him from the collection of neurons available for a future utilization. The following paragraph gives a very short and simplified description of the performance of the system.

An input binary vector I is submitted to the system. To this vector corresponds, after competition between the neurons of the hidden layer, one winning neuron j. This winner j is regarded by the network as the best representative of the input vector I. The neuron j, through its back connections, generates a binary output vector O on the comparison layer. O and I are compared in the specialized neuron (see above). (Remember that all components of O and I take the values 0 or 1). If the ratio of the number of the components of to the number of components of the same value for the input I is above a threshold value that has been fixed, the neuron j is confirmed as the representative of the class of I. The weights of the connections between j and I are modified to increase the strength of their association. On the contrary if the difference between O and I is too large, the process restarts with the neurons of the hidden layer less the former winner. If all the neurons of the hidden layer have been reviewed without success, a new hidden neuron is added. This new neuron will be the representative of the class of I that is an input coming out of an unknown new class. The weights of the connections of that new neuron are fixed by the same procedure as for learning. The choice of the threshold value is of great consequence on the number of groups delimited by the network. As for the Kohonen net the learning process intends to rotate the weights vector of the backward connections towards the input vector. The adaptation of the structure to unknown inputs by the inclusion of new hidden neurons allows solving the dilemma of stability versus plasticity.

ART2 are ARTNN that are capable of dealing with analog signal transitions; ART3 is an extended version modeling the action of chemical transmitters of the nervous system. ARTNN have had a mitigated success. Results are not stables and change with the order in which the inputs are offered to the network. ART nets are very sensitive to noise. It is difficult to find the good parameters and the cost of computing is high. Genetic algorithms, genetic programming and particles swarms have been used to facilitate the design or improve the performance of ANN.

REFERENCES

[1]   Eccles JC. The Evolution of the Brain, Creation of the Self. London: Routledge ed. 1989; p. 257.

[2]   Shannon C. A mathematical theory of communication. Bell Syst Tech J 1948; 27(3): 379-423.

[3]   Turring A. On computable numbers with an application to the Entscheidung problem. Proc Lond Math Soc 1937; 42: 230-65.

[4]   Mc Culloch W, Pitts W. A logical calculus of the ideas immanent in nervous activity. Bull Math Biophys 1943; 5: 115-33.

[5]   Hopfield JJ. Neural networks and physical systems with emergent collective computational abilities. Proc Natl Acad Sci USA 1982; 79(8): 2554-8.

[6]   Rosenblatt F. The Perceptron, a perceiving and recognizing automaton. Cornell Aeronautical Laboratory. Report 85-460-1, 1957.

[7]   Widrow B. Generalization and Information Storage in Networks of Adaline "Neurons". In: Self-Organizing Sytems. Symposium proceedings: Youtiz M, Jacobi G and Goldstein G eds. Spartan Books, Washington DC 1962; 435-461.

[8]   Hebb DO. The Organization of Behavior. A Neuropsychological Theory. Wiley ed. New-York NJ, 1949; p. 317.

[9]   Amit D, Gutfreund H, Sompolinsky H. Information storage in neural networks with low levels of activity. Phys Rev A 1987; 35(5): 2293-303.

[10]   Bruck J. On the convergence properties of the Hopfield model. Proc IEEE 1990; 78 (10): 1579-85.

[11]   Ackley D. H, Ed. A connectionnist machine for genetic hillclimbing. Boston: Kluwer Academic Ed. 1987; p. 212.

[12]   Saarinen J, Kohonen T. Self-Organized formation of colour maps in a model cortex. Perception 1985; 14(6): 711-9.

CHAPTER 3

Sensors, Behaviors, Muscles

Abstract: External and endogenous phenomena are detected by sensors. Most of them are just modified neurons. Some others (in human retina, in ears) are hypersensitive cells. Sensorial data are processed in the brain which send orders to the muscles. From the signals of a great number of olfactory receptors, brain extracts the ordered list of the six more strongly excited. Man can learn to recognize $2 \cdot 10^3$ smells (and dog 10^6). Eyes of bees or frogs have poor keenness sensitivity. Human's retina has very large keenness and sensitivity. In frog, visual signals are processed on the retina. In man, they are sent to the brain and processed by several parallel paths. Adaptable software governs inborn or learned behaviors (for instance walking). Only one out of several competitive behaviors (for instance the various possible directions of eyes) has to be chosen. Choosing centers (for instance superior colliculus) select the more strongly excited among several competitive instinctive drives, which are sophisticated combinations of several sensorial signals. Behaviors control the muscular activity. They are simple automatisms in frogs. In men, they consist in layers of control devices. The lower-level unit is based on two antagonist muscles and their embedded sensors. This unit governs muscular tonus. The cerebellum is in charge of the phasic excitations. Various signals combine to build a representation of the body position.

Keywords: Behaviors, body representation, cerebellum, choices, colliculus, competitive instincts, eye, hearing, instinctive drives, muscular embedded sensors, muscular tonus, optical coding, phasic excitations, pleasure impulse, retina, sense of smell, sensors, striated muscles, thirst control, visual pathways.

3.1. Sensorial Analysis

Introduction: All sorts of sensors are known: they respond to light, sound, chemical molecules, acceleration, magnetic field in pigeons, electric field in sharks. We have also to deal with endogenous sensors that respond to osmotic pressure, joint angular position, muscular fibre activity. In all such cases, the exciting phenomenon induces the opening of post-synaptic pores. The cells supporting these pores are usual neurons or, in the retina and the inner ear, hypersensitive and energetically expensive pseudo-neurons

3.1.1. The Sense of Smell

Main results: Description of the successive operations leading to the recognition of learned smells.

3.1.1.1. Pheromones and Smells

Olfaction is the most archaic sense. The inputs are perfumed molecules brought by the air. They act on sensors composed of normal neurons with specialized chemo-transducers in place of the usual post-synaptic receptors. These receptors release messengers, which slowly open ionic channels. The outputs are a) recognized smells, each of them triggering a specific behavior and b) an intensity measurement used for guiding the behaviors.

The use of pheromones by insects is very simple. The device is sensitive to a defined set of perfumed molecules: the pheromones. Each pheromone is detected by only one type of receptors. The receptors act in a linear way allowing intensity measurements. The enamored male butterfly uses it to find the direction of the female butterfly. When excited, each receptor triggers an all-or-none inborn behavior.

The smelling device of mammals is awfully more complicated because it deals with an undefined set of perfumed molecules. (In an analogous way, the immune system deals with an undefined set of attacks). A systematic genetic study of the receptors [1] and quantitative histology lead to understand how it works. The device is based on three parts: a) the receptors in the nose mucosa; b) the mitral cells and the other cells of the olfactory bulb; c) the olfactory cortex.

Olfaction is much more useful to dog than to man. In dog, identification of another dog or of a human is mainly due to the smell sense; recognition of a smell triggers a behavior. In man, identification is mainly visual; recognition of a smell induces often memory searching, it seldom triggers a behavior. It seems from experimental results that a dog recognizes about $N = 10^6$ smells and a man only $N = 2.10^3$.

3.1.1.2. The Receptors

All the mammals own the same family of about 1000 olfactory specialized genes. But only a part of them is functional. Each functional gene codes for a type of receptors. Thus, man owns about 350 types of receptors, dog about 700. The olfactory epithelium is divided in 4 zones [2].

The total number of receptors increases from the mouse to the dog and briskly decreases for man: 2.10^6 receptors in a mouse; 6.10^7 in a rabbit; 10^9 in a dog; 5.10^7 in a man. Thus, the number of receptors of a type is $10^9/700 = 1.4 \ 10^6$ in a dog and only $1.4 \ 10^5$ in a man. High numbers are related to sensitivity and accuracy of intensity measurements.

Receptors of a type are excited by many perfumed molecules. A perfumed molecule excites receptors of several types. The excited neurons have a linear behavior: when excited by the same molecule, receptors of different types induce different firing rates.

3.1.1.3. The Olfactory Bulb

The olfactory nerve fibres arising from several receptors of the same type converge through a relay on the mitral cells, which, as the epithelium, are divided in four zones. (They converge also on the tufted cells which inhibit some far away mitral cells, acting as a selecting device).

The number of mitral cells is 6.10^4 for the rabbit, 10^6 for the dog and 5.10^4 for the man. They are transient swift neurons forming such a tight bundle that, when unexcited, their spontaneous activity is synchronized and amplified by mutual electric coupling (as in α EEG waves, see Part I, Appendix M), generating intrinsic waves with a frequency 70 Hz.

The collected receptor data enters into a progressive recruitment device (the mitral cells), which builds a logarithmic measurement of each type of perfumed molecules concentrations on the nose mucosa. They are 1700 mitral cells per receptor type in dog and 140 in man. Their transmission capacity is limited by noise (see 1.3.5.1), each of them is able to distinguish about 200 messages. Thus, for each kind of receptor, the olfactory bulb is able to distinguish about 3.10^5 logarithmic intensities in dog and 3.10^4 in man.

Let us call A, B, C... the different receptor types. The main output of the olfactory bulb is the list $\{I_A, I_B, I_C...\}$ of the different logarithmic intensities. A secondary output is the global intensity, which is used for orientation.

3.1.1.4. The Olfactory Cortex

The analysis of smells is supported by the association of a choosing device and a filtering one.

3.1.1.4.1. The Simplest Scheme

The Choosing Device: It seems to be a f = 6 floors device. Its inputs are the list $\{I_A, I_B, I_C...\}$ of the different intensities. We know the scheme (see 1.3.4.4) of a competitive modulating device choosing, among several inputs, the one of strongest intensity. Its output is an all-or-none signal, the meaning of which is "A is the more excited type of receptor". This information is not sufficient to characterize smells. The olfactory six floors choosing device builds the classified list of the six greatest intensities, for instance A first, then B, then C... The device uses six parallel and similar elementary competitive modulating devices with a slight change: the output A of the first device is used as a signal strongly inhibiting the input I_A of the second device, which then chooses the second more intensive input I_B.

The output of the choosing device is then the ordered list {A, B...F} of the names of the six more excited types of receptors. (At another time, another signal will dominate and will be linked with another list). This output is a 6 letters word (each letter used only once) written with an alphabet of n = 350 letters for man and n = 700 for dog (as many letters as receptor types).

The Filtering Structure: The number of columns in each floor is equal to the number N of recognizable smells (10^6 in dog and only 2.10^3 in man). After learning, each of the N pathways is associated with a 6 letters password, for instance A*, B*, C*... When a message coincides with the password of some pathway, the message is transmitted and the output of the pathway fires.

The structure of the filtering device (the number of columns and the number of floors) is inborn. But the passwords are learned. Part IV is devoted to the study of learning. We will describe then the neuronal circuitry and understand how it works. Although one can learn only a finite number of passwords (10^6 for dog), the number of possible messages or possible passwords (the number of six letters words written with a huge alphabet) is almost infinite. So, we explain that dog can be exerted to recognize the most unexpected smells (such as the smells of drug, explosive or truffle).

A Fragmented Scheme: We note that the simplest scheme applied to the dog would require several times f.n.N neurons ($6*700*10^6 \approx 4.10^9$ neurons), a huge number, much greater than allowed by anatomy and much greater than the optimistic Hopfield's limit which is then $1.4 \ 10^5$. Then, we remember a still unused observation: the olfactory epithelium and the olfactory bulb are divided in

4 zones. Starting from this experimental fact, we are led to look at a fragmented device. We assume that four separate partial devices select "letters" from 4 different alphabets, each of them made of n/4 = 175 "letters" for dog. Each of these partial devices recognizes "words" made of φ different letters taken in the partial device alphabet. The total device recognizes "sentences" made of 4 "words" (one from each partial device). It is able to recognize $N=10^6$ "sentences". Then, each partial has to recognize $N^{1/4} \approx 30$ outputs. On the top of the partial devices, a double combinatorial net making AND operations recognizes two words sentences: it associates each recognized word of the first alphabet with each recognized word of the second one and each recognized word of the third alphabet with each recognized word of the fourth. Another combinatorial device associates two word sequences to recognize four words sequences. Two floors are needed by the combinatorial devices. Thus, if the total device owns f=6 floors, the partial devices have at their disposal φ= f-2 = 4 floors. To summarize: the passwords are sentences of 4 words of 4 letters. Floors 1 to 4 require few neurons. The fifth floor is made of two devices, each with 30+30 = 60 inputs and 30*30 $\approx$ 10^3 outputs (one for each node). The sixth floor (the biggest) owns 10^3+10^3 inputs and 10^6 nodes exciting 10^6 outputs. Thus, the required number of neurons is realistic and is not excessively far from the theoretical Hopfield's minimum that has been above mentioned.

Subjective Choices: Feedbacks coming from non-olfactory parts of the brain can modulate some inputs of the choosing box, for instance I_A. Then, A will be chosen (as long its intensity is not zero) even if its intensity is not the greatest: a dog following a track is not distracted by strong parasitic smells.

3.1.1.4.2. Remark

The olfactory cortex uses only ratio of molecules concentrations (differences of logarithmic intensities): the analysis does not depend on the global intensity; it is invariant under a global change of intensity (under a change of the distance of the perfumed source). The global intensity is used by a separated pathway. Building of invariant signals and separation of pathways will be fundamental principles in the following Sections.

Note that the classification of intensities does not work if the receptors are saturated: excessively condensed perfumes cannot be analyzed.

3.1.2. Vision I: The Retina

Main results: Peculiar properties of cones and rods, optimal coding

3.1.2.1. Introduction

While most of sensors are usual neurons with a modified receptor, the eye and the ear (Section 3.1.4) are very special devices. The leech owns 5 photo-sensors allowing some reactions to the surrounding luminosity. But insects and all the vertebrates own eyes that are an optical apparatus forming an image on a mosaic of photo sensitive cells. The retinas of all the animals use almost the same molecular components.

The behavioral usefulness of vision depends mainly of the angular resolution and on the sensitivity of the retina. The problem is that any change increasing the resolution (for instance using a diaphragm) tends to decrease the sensitivity and that the user needs of both high sensitivity and good resolution. Thus, the evolution of eyes appears as a sequence of more and more subtle optimizations to adapt the eye structure to a general behavior particular to each animal.

3.1.2.2. The Performances

3.1.2.2.1. Sensitivity

Animals have to deal with the following typical input illuminations (Table **3.1**). The illumination unit is candela per square meter cd/m^2.

Table 3.1. Typical illuminations.

Pitch-dark night	10^{-5}
White sheet of paper in pitch-dark night	10^{-4}
White sheet of paper under a lamp	50
White sheet of paper under bright sun	$2\ 10^4$
Looking at the sun	$2\ 10^5$

We have to compare these illuminations with the detection ranges of various animals. The range of mammals is extended by the use of a double system with two types of sensors. Central vision uses cones. It is well adapted to a very sharp analysis of the shapes and the colors of motionless or slowly moving objects in day light. Peripheral vision uses rods. Rods are well adapted to a coarser analysis in a wider visual field and both in dim or day light. They allow also the detection of fast moving objects. Table **3.2** gives the ranges and the dynamics (ratio of the saturation to the threshold illuminations) for men and honeybees.

Table 3.2. Illuminations seen by men and insects.

	Man's Rod	Man's Cone	Honeybee
Threshold	$1.7 \ 10^{-4}$	$7 \ 10^{-3}$	$3 \ 10^{2} \, cd/m^2$
Saturation	$5 \ 10^{2}$	$2.5 \ 10^{4}$	$3 \ 10^{5} \, cd/m^2$
Dynamics	$3 \ 10^{6}$	$3.6 \ 10^{6}$	10^{3}

Honeybee, drosophila and other insects have almost the same ranges. With such a high threshold, insects see only a lamp filament. Cat, dogs and man have the same dynamics, three thousand times larger than the insect dynamic.

The rods and the cones of cat and the cones of dog are slightly more sensitive than the human ones. The thresholds are: cat's rod **5 10** $^{-5}$; cat's cone **2.5 10** $^{-3}$; dog's cone **10** $^{-3}$ cd/m^2.

3.1.2.2.2. Angular Resolution

In optician's language, to have a 7/10 vision means to have an angular resolution equal to 10/7 of a one minute angle. It is more convenient to use radians: 1' corresponds to 3 10 $^{-4}$ radians, that is the ability to discriminate objects of 3 mm at 10 m of the observer.

Here are the angular resolutions of some typical animals: Drosophila **0.12** rad; honeybee **0.04**; Tortoise and frog 0.03; dog **2.5 10** $^{-3}$; cat and man (with a good sight 10/10) **2 10** $^{-4}$ rad.

These numbers apply only to the central vision and only if I > 10 cd/m^2 (for man). For weaker illuminations, we will see that gap junctions cause a decrease by a factor 20 of the angular resolution. On the other part, the lateral vision is less acute than the central one.

3.1.2.2.3. The Compromise Between Sensitivity and Resolution

The image on the retina of a punctual luminous source is a little spot, the diameter of which depends diffractive phenomena and on the aberrations due to the lens. The resolution cannot be smaller than this diameter. But it can be greater: wide diameter photo-receptors have the advantage to collect more light than those with small diameters.

For little size animals (as insects) for which the eye size is not much larger than the wave lengths of the light, the best optical solution is a compound eye made of a great number of nearby independent lenses, the <u>ommatidies</u>. The eye of the honeybee is close to the best diffractive optimum with 5000 ommatidies (resolution: 2.5 d° = 4 cm à 1 m) while the fly is far from it with wide receptors and only 800 ommatidies (resolution: 7 d° = 12 cm à 1 m).

For tortoise and frog, the diffractive limit would be about 10' (3mm at 1 m). But much wider receptors are needed to collect a sufficient light energy for normal diurnal illuminations. They allow only a 2 d° (3 cm at 1 m) resolution.

Cats or men use at best their optical apparatus. But their 1' resolution implies very small photo-detectors areas, hence a weak signal which has to be processed through a hypersensitive cell body.

Although the properties of the dog visual system depend strongly on its breed, we present some average results [3]. Dog owns only 17.10^4 ganglion cells (their axons are the fibres of the optical nerve) when man owns some 10^6 of them. Dog' retina does not own red pigments, its fovea contains a great ratio of rods (90% when it is 0% in human fovea). The result of this adaptation is acuity six times smaller in dog than in man. This bad performance allows an enlargement of the cornea diameter, which increases the retina illumination, but increases the geometrical aberrations of the crystalline lens and produces a blurring of the pictures incompatible with human acuity, but compatible with the poor acuity of the dog (a 3/10 vision).

3.1.2.2.4. Behavioral Consequences

From their dance, honeybees learn the direction and the distance of flowers. Their eyes are sensitive to the polarization of light and so can distinguish the direction of sun. This data and an internal clock allow navigating from the beehive to a rough location of flowers. Their sight is able to recognize a flower by its color and the number of its petals. Their angular resolution allows them to identify a little flower (about 1 cm) at less than 20 cm of it. So, vision can be used only for the final phase of gathering. The intermediate phase (between navigation and gathering) is not governed by vision, but by olfaction.

Wolves and many dogs are night hunters: while human vision is optimized for daylight, dog sees almost as poorly in day that in night and uses only a little part of its brain to analyze the visual data. Olfaction and hearing are its leading senses, vision acts as a complement. Only men, cats and probably monkeys have a fully visual behavior.

3.1.2.3. The Hypersensitive Visual Cells

The main result of the above description is the appearance of a very high dynamics in the human eye. This property is due to the use of hypersensitive cells as sensors. They are made of the same molecular components as usual neurons, but their control potential depends in such a unusual way of the post-synaptic channels opening that their signal is proportional to the logarithm of the illumination (with a very great sensitivity for weak illumination).

3.1.2.3.1. From the Input Illumination to the Opening or Closure of Post-Synaptic Channels

For all the animals, the light induces the activation of photo-sensitive molecules, the opsines, bound to specific pigments. In man (but not in honeybee), the photonic flow on the retina is modulated by the closure of the pupil (Appendix A). Photon activated opsines are unstable molecules which quickly decay into an inactivated state. Inactivated molecules slowly return to their initial state. After a strong illumination, the slow readjustment to darkness in man takes 7 minutes for rods and 2 minutes for cones. The result of this phenomenon (named bleaching) is a loss of the receptors sensitivity under a illumination.

3.1.2.3.2. A Comparison Between Honeybee Sensor, Human Rod and Human Cone

In the three of them, the one photon absorption by an opsine molecule activates a specialized G protein which, directly or indirectly, acts on the opening or the closure of post-synaptic channels.

3.1.2.3.3. The Honeybee Sensor

It is a usual neuron with a modified receptor. The weakest detected illumination induces a control potential equal to the noise 0.15 mV. The maximum control potential is 32 mV. Thus, if the potential was strictly proportional to the illumination, the dynamics would be 32/0.15 = 200. Non linear effects (due to the summing of micro-potentials and to the bleaching) increase the dynamics by a factor 5. The parameters of the photo sensitive part (diameter, opsine concentration, decay time of the activated state) are optimized in such a way that the threshold illumination is adapted to the behavioral needs.

3.1.2.3.4. The Human Rods (Fig. **3.1**)

They are hypersensitive cells made of a transducer, an internal segment and a pre synaptic ending. The transducer membrane contains Na^+ channels which *close* under the effect of illumination. The membrane of the internal segment contains some usual passive channels and active Na^+ pumps.

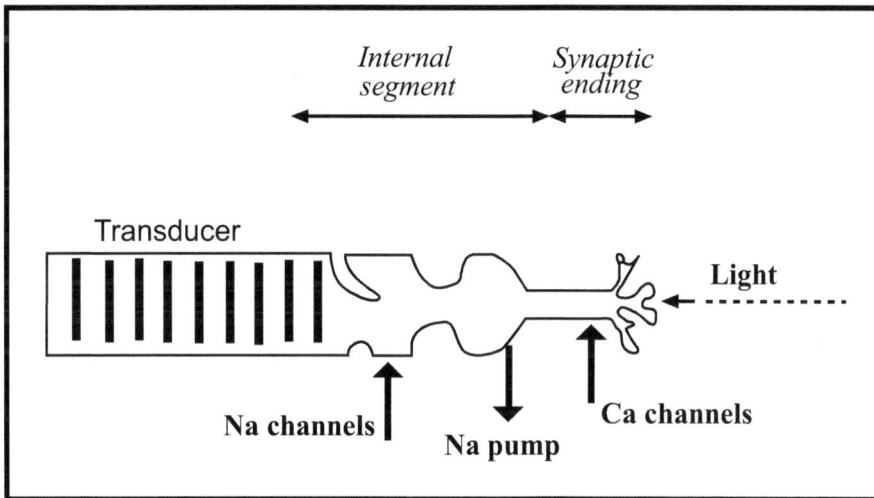

Fig. (3.1). Human rod sensor.

When all the transducer Na channels are closed (strong illumination), the active pumps are at rest, the cell evolves towards the usual ionic concentrations and towards the usual rest potential $V_R \approx$ -70 mV. In darkness, the channels are open, the active pumps are working; the balance between the passive Na current and the active pumping imposes a rest potential close to -40 mV. The synaptic ending contains calcium channels, the opening of which depends on the internal segment potential. Ca^{2+} ions entrance induces the release of glutamate molecules which excites the following retina layer.

As in the honeybee sensor, the parameters of the photo sensitive part are optimized in such a way that the cell potential at threshold differs from the potential in darkness by 0.15 mV. The potential in darkness and at saturation are -40 and -70 mV. If all the phenomena were linear, the dynamics would be (70-40)/0.15 = 200, as in the honeybee sensor; the measured dynamics is more than 10^6. Such a value implies that a violently non linear phenomenon occurs for weak illuminations: it must be such that dV/dI is infinite when I = 0. This phenomenon is probably due to the reversible binding of one activated opsine with two Na channels (see Appendix B): the potential is roughly proportional to Log(I) and not to I, giving thus a high weight to weak illuminations and extending the dynamics.

Remark: The sensitivity of the cell depends on the ratio between the number of opsine molecules and the number of Na channels, which has to be high enough. Thus we explain that the transducer length is greater than the internal segment.

The human cones: They use the same mechanism that the rods. They exhibit the same potential in darkness (-40 mV): the ratio between the number of pumps and the number of channels is the same in rods and cones (see Appendix B). The illumination at saturation is higher: either the lifetime τ of the activated opsine or the ratio between the number of opsine molecules and the number of channels are weaker in cones. On the other hand, to be usable in semi-darkness, the cones are able to work below the noise level: we have seen in part I how gap junctions allow sacrificing resolution to sensitivity.

The Cost of Hypersensitivity: In most of cases, usual neurons are more useful than hypersensitive cells, which suffer from several disadvantages:

a)   The output signal of the cones and the rods is the release (or, more accurately, the variation of the release) of a transmitter, the glutamate. This release depends on the opening of voltage dependant calcium channels at the end of the cell body. But the cells are optimized to use a weak number of Na channels, which can act only on a weak number of synaptic endings, the glutamate release is too weak to excite a usual neuron. The retina needs amplifying cells between the cones and the optic nerve.

b)　In most of cases, the logarithmic response of the hypersensitive cells is less convenient for processing than a linear one: the output of the rods and cones have to be used as inputs for some usual neurons.

c)　While usual neurons generate spikes (an almost digital behavior), hypersensitive cells exhibits an easily jammed analogical behavior.

d)　Hypersensitivity results from active pumping. While the metabolic expenses of usual neurons are weak when the cell is silent, the expenses of hypersensitive cells are permanently high.

3.1.2.4. The Structure of the Retina

Since the first works of Hartline [4], Granit [5] and Kuffler [6], the detailed circuitry of this intricate system is now fairly well elucidated. The retina is made of 3 layers, the photo sensors, the bipolar cells and the ganglion cells. The axons of the ganglion cells make up the optic nerve. Two additional layers (horizontal and amacrin cells, Fig. **3.2**) assume long distance connections inside the retina.

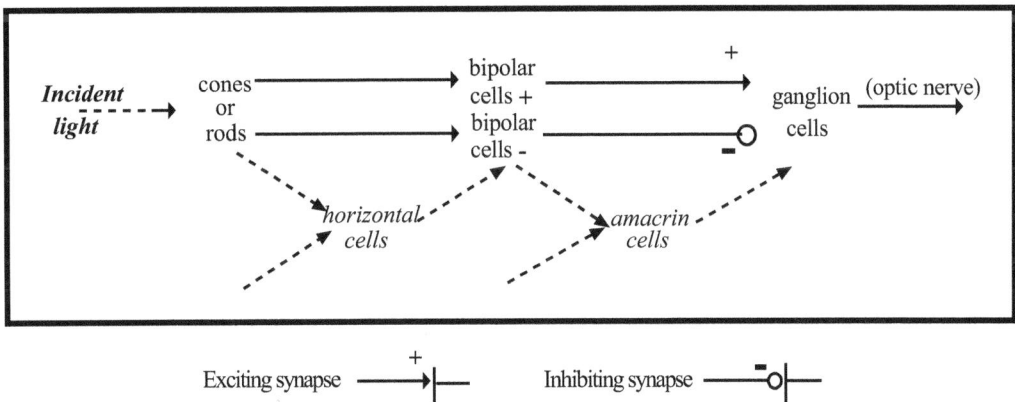

Fig. (3.2). Connections between layers of the retina.

3.1.2.4.1. The Global Distribution of Sensors and Ganglion Cells

The central part of the human retina (the fovea) is 4° wide. It contains a great density of cones, but not any rod. The high density of cones in the fovea allows a very good central acuity (about one minute of arc for a normal eye) while the lateral acuity is greater than one degree. 4.10^4 cones excite 4.10^4 ganglion cells through 12.10^4 bipolar cells. On the contrary, the fovea of dogs contains only 10% of green and yellow cones and 90% of rods.

The peripheral retina of each eye is 170° wide (140° binocular + 30° monocular). It contains a smaller density of cones and a huge number of rods: 125.10^6 rods and 5.10^5 cones excite 10^6 ganglion cells through 6.10^6 bipolar cells (half of them linked to cones).

3.1.2.4.2. The Components

Cones and Rods: The photosensitive molecules are specific proteins, the opsines, bound to specific pigments. The absorption spectrum of the rod receptor (the rhodopsine) is maximum for green ($\lambda = 496$ nm). Its sensitivity spreads from blue to yellow. The spectra of the three cone receptors have the same shape, but with their maximum translated to blue (419 nm), green (531 nm) and yellow (559 nm). Processing of the signal for three slightly different spectra leads to the colour sensation. Some animals (as the bees) own ultra-violet sensitive pigments. Dog cones own only green and yellow pigments.

Rods and cones release constantly glutamate. Light hyperpolarizes them, the release of transmitter decreases. They excite directly horizontal and bipolar cells.

Intermediate Cells: All of them are hypersensitive cells amplifying the weak signals delivered by rods and cones. They exhibit linear input/output relations.

One knows today 4 types of horizontal cells, 9 types of bipolar cells, 14 to 20 types of amacrine cells and 8 to 20 types of ganglion cells.

Horizontal cells receive feed backs from bipolar cells and modulate the sensor activity. In particular, they govern the gap junctions between rods, which allow avoiding noise in dim light.

Bipolar cells act directly or through amacrine cells on the ganglion cells. Several transmitters are used (glycine, GABA). Half of the bipolar cells (Bi+) are exciting, the other are inhibiting (Bi-). In day light, one rod bipolar cell excites one ganglion cell. In dim light, 30 rod bipolar cells excite one ganglion cell through one amacrine type II cell.

Ganglion Cells: The axons of these ordinary neurons make up the optic nerve. They are swift transient neurons (see Chapter 1). Bipolar cells are in charge of their threshold modulation. Their initial firing rate F_0 is about 1200 s^{-1}, their asymptotic firing rate is around some 50 s^{-1}. It is reached after a characteristic transient time T_D ($T_D \sim 0.01$ s). Ganglion cells are excited and inhibited by the bipolar cells and by a threshold modulation. When in darkness or in diffuse light, the ganglion cells are driven by the modulation with a mean firing rate about 4 s^{-1}.

3.1.2.4.3. The Receptive Fields

Definition: The linearity of all the coupling implies that the excitation of any ganglion cell is a weighted sum of the outputs of all the rods and cones (which are proportional to the logarithm of the illuminations). To obtain an evaluation of the global illumination, the weights have to be same for all the photoreceptors. On the contrary, to keep a good angular resolution, the weights have to differ from zero in only a little area of the retina. This area is called the receptive field of the ganglion cell. In fact, we look at excitatory receptive fields [+] and inhibitory ones [-]. The shape of these areas can be circular or elongated. Often, the inhibitory field surrounds the excitatory one or the contrary.

Balanced Receptive Fields: Receptive fields are **balanced** if, for any diffuse luminosity, the contributions of the [+] and the [-] fields are equal but of opposite signs. Thus, a ganglion cell excited by balanced fields is almost mute for any diffuse luminosity. The ganglion cell fires if the receptive exciting field is more enlightened that the inhibiting field and its firing rate increases with the difference of luminous intensities.

Self-Healing Properties of Balanced Fields: When a bipolar cell or more often a photo-sensor is destroyed, the receptive fields stop to be accurately balanced and we would expect the ganglion cell to fire under diffuse light. But the threshold modulation restores the rest excitation just above the threshold. Thus, the only observable change is a slight decrease of the firing rate *versus* spot light intensity.

3.1.2.4.4. Time Behavior of Ganglion Cells

Sustained and Brisk Behaviors: In a great lot of experimental works, one enlightened suddenly a little spot of the fovea while an electrode measured the firing of a ganglion cell. If the spot acts on the exciting field of the cell and not on its inhibiting one, we observe an ON sustained behavior: a short burst of high firing rate spikes is followed in many cells by a sustained activity with a firing rate in the range 30 to 60 s^{-1} (Fig. **3.3**, left). In 10% of the ganglion cells, one observes an ON brisk behavior (Fig. **3.3**, right): the cell is mute after a burst of about 10 spikes. (This behavior is due to the synaptic delays of the bipolar cells, see the Appendix C).

Now, if the spot acts on the inhibiting field, the neuron is mute, the special K channels (see part I) close. When the light is cut out, the threshold modulation $U_{Mod} = U_{Thr} + 0.15$ mV generates a burst (OFF behavior) with the firing rate F_0 (before transient effect) $F0 \approx 500$ s^{-1}. Note that in some animals as the mouse and the rabbit, one observes sometimes bursts during a short time at the beginning and at the end of the enlightening (ON-OFF behavior).

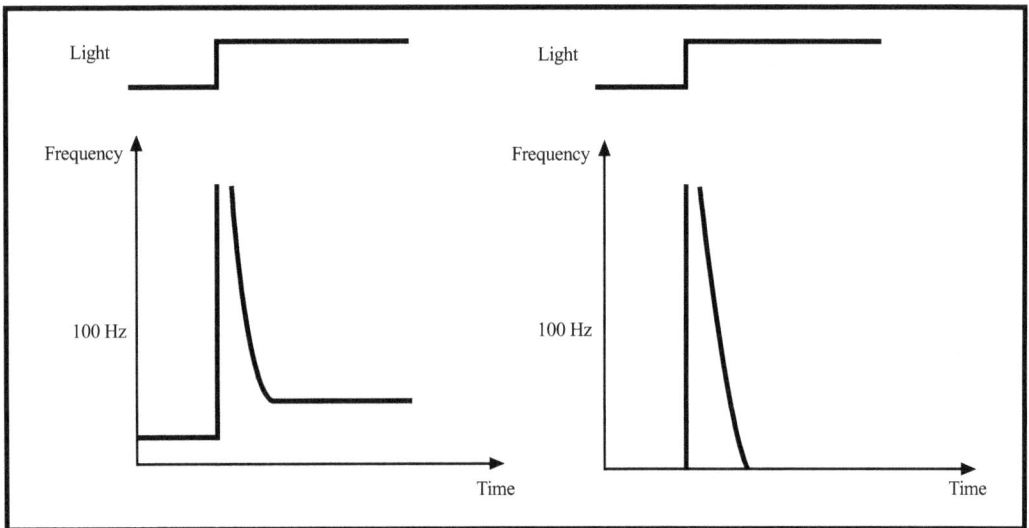

Fig. (3.3). Sustained or non-sustained responses of a ganglion cell to constant illumination.

Transient Ganglion Cells are Convenient to Detect Motion: We look at transient ganglion cells with balanced receptive fields [+] inside and enlightened by a moving spot (diameter d, velocity on the retina v). If d is very small, the response of the ganglion cell is weak. If the spot is wide enough to cover both the [+] and [-] receptive fields, the cell does not respond. The response is obviously maxima if the diameter of the spot is equal to the diameter of the [+] field. In the same way, if T_D is the characteristic time of a transient response, we observe that the response is maxima if $v.T_D$ is equal to the diameter of the inside [+] field.

For human cones, ($T_D = 10^{-2}$ s and a 1' receptive field) the critical velocity is a target displacement of 45° in 0.5 minute, a very slow motion. For peripheral rods (same T_D, 1° receptive field), the critical velocity becomes a target displacement of 45° in 0.5 second, a very fast motion.

3.1.2.5. The Visual Coding

How are the receptors connected to the ganglion cells?

3.1.2.5.1. Specialized or Optimal Coding

***An Example of Specialized Coding: The Frog Retina* [7]:** The frog, a primitive vertebrate, recognizes three classes of objects, worm type, fly type and flying predator type such as vultures. Electrophysiology shows that the signal recognizing is made at the retina level: the retina contains neurons specialized for vultures, flies and worms. When one of these neurons is excited, it triggers a stereotyped behavior. These neurons are transient ganglion cells with balanced receptive fields [+] inside. The size of the receptive field and the characteristic time T_D are adapted to the target characteristics.

Color Coding: We have to deal with the three types of cones (green, red, blue) and with three associations at the bipolar cells level: crimson = blue + green, magenta = red + blue, yellow = red + green (and obviously green + blue + red = white). At the ganglion cells level, six combinations are built up from these colors: center green with red background, center green with magenta background, center red with green background, center red with crimson background, center yellow with blue background, center blue with yellow background.

Optimal Coding (see Appendix D): Signal recognizing is made by the cortex. The function of the ganglion cells is to code and transmit the data without information losses and with a maximum of balanced receptive fields. But all the receptive fields cannot be balanced: global illumination can be evaluated only by unbalanced fields. More generally, one shows that to transmit the values of the intensities reaching N black or white sensors, one needs N/4 cells with balanced and round fields, N/2 cells with balanced, but elongated fields and N/4 cells with round but unbalanced fields. Dealing now with the 6 types of color coded fields we have to replace 1/(3+1) cases with balanced and round fields by 6/(3+6) cases. The above results become: 6N/9 (67%) balanced, chromatic and round receptive fields, N/9 (11%) black and white balanced elongated fields, 2N/9 (22%) black and white, round but unbalanced fields. These theoretical results have to be compared with the 65, 11 and 24% observed in macaque.

Real Coding in Men: For each type of balanced, chromatic and round receptive fields, the numbers of ganglion cells with [+] field and [-] field inside are equal. The distribution of unbalanced black and white cells shows a systematic symmetry: half of them are silent in darkness while the other half are spontaneously firing in darkness with a firing rate about 100 s^{-1} and their firing rate decreases for strong light intensity (black is handled as a color).

Real Coding in Some Animals: The frog owns a completely specialized coding. The rabbit [8] owns a double system: 27% of its ganglion cells generate specific alert signals: they are direction and motion selective. In addition to an optimal coding, the cat owns a set of special ganglion cells (the W cells) with a wide unbalanced [+] field. These cells, as numerous as the sustained ones, seem to be adapted to the analysis of shapes and colors during night.

3.1.2.5.2. Quantitative Schematization of the Retina Interconnections

We have to build the matrix linking the cone outputs (the logarithms of the illuminations) to the ganglion cells steady frequencies (Appendix E). This intricate matrix takes into account a) the progressive recruitment of 10 neurons with increasing illuminations b) the 6 color combinations and the 9 fields types in an optimal coding c) the black-white symmetry.

3.1.3. Vision II: The Brain Visual Analysis

Main results: The interaction between objective and subjective colliculus choices.

3.1.3.1. The Experimental Knowledge

The input of the device is a continuous flow of data transmitted by the optic nerve (made of the axons of the ganglion cells) after a very intricate retinal coding. The outputs are behaviors triggered by the recognition of various objects.

3.1.3.1.1. The Experimental Methods

To follow the course of visual signals is a very difficult task because:

- The number of neurons that are involved in the propagation of such signals is huge.

- The number of branching and the progressive dilution of the circuitry is large.

More over visual stimuli induce motor and mental reactions without direct links with the visual system.

Experimenters have to their disposal four instrumental methods:

a) *In vitro* studies of histological sections. These studies give some ideas about the connections and the main cortical pathways; it gives also some ideas about the number of neurons in a cortical area.

b) *In vivo* measurements of the electric activity of an accurately located neuron when the eye is excited by various luminous signals. Results for men are deduced from experiments on primates. This method gave a huge number of results about the retina, many results about the first cortical area (Brodman's area 17), some results about the colliculus and the lateral geniculate nuclei (LGN), practically no results about the other parts of the device.

c) Imagery of the working cortex. Activity of a cortical area induces a local change in the blood oxygenation which is measurable by magnetic resonant imagery (functional MRI, PET scan). It gives a rough idea of the propagation of the brain activity from its occipital parts (area 17) toward the frontal parts.

d) Examination of the effects of various human lesions.

e) To the results of these methods we have to add the results of obvious observations with special attention being given to the following points: we recognize an object with or without binocular vision, whatever are its distance, its illumination and often its orientation; we recognize black and white pictures and simple sketches of its contours.

3.1.3.1.2. The Main Thalamic Relays

The Right and Left Lateral Geniculate Nuclei: They are made of 6 layers; each of them is a map of a hemi visual (right or left) field of both eyes. In the right LGN, signals from the right eye excite layers 1, 3 and 5, from the left eye layers 2, 4 and 6 (Fig. **3.4**); they are transmitted to the right hemi-cortex. Brisk ganglion cells excite layers 1 and 2, sustained cells the others. Each layer is a map of the retina. More: co-ordinates $[X_1, Y1]$, $[X_2, Y2]$ and so on of points of the six layers associated to the same point x,y of the retina are lined up. The LGN are in charge of the separation of the sustained and the transient (brisk) signals, they inhibit the transmission toward the cortex during sleep and during rapid eyes motion and they take part in the making of 3D perception. Note that many mammals do not own Stereoscopic vision: in man, monkey and cat, the axis of both eyes are nearly parallel. The binocular visual field (140° wide) is extended on each side by a monocular field 30° wide. In rabbit, eyes are on the side of the head. The divergence between axes is about 160°: the rabbit does not benefit of binocular vision.

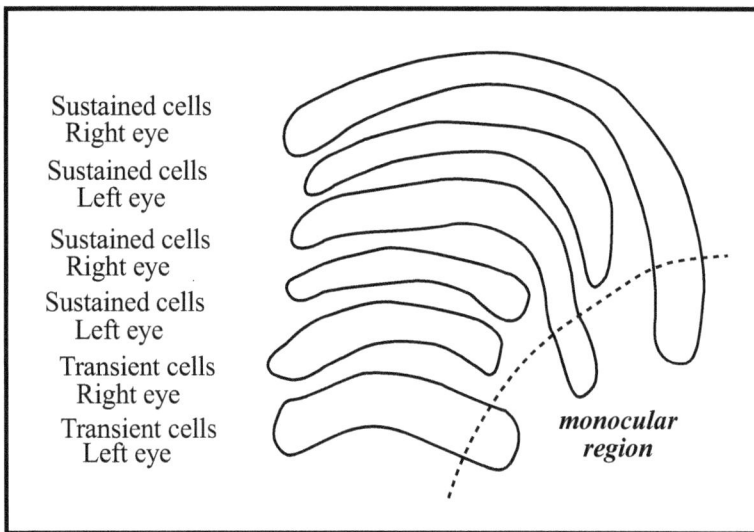

Fig. (3.4). Functional structure of the geniculate nuclei.

The Superior Colliculus: It is in charge of the competitive visual choices. So, it is strongly connected with the control of eyes direction and head position. It contains a retina map exhibiting a remarkable property: an artificial stimulation of the point X,Y of the colliculus associated with the point x_{before}, y_{before} of the retina induces a **saccade** (a brisk eye motion) bringing the image of the target at the center of the retina.

The very close inferior colliculus is in charge of the aural alert signals (see Section 3.1.4). So, we understand why we turn the head toward the direction of a shout.

The pretectum: It is in charge of the control of the pupil diameter.

3.1.3.1.3. The Visual Pathways

The Control Pathways: Some ganglion cells, which measure the global illumination of the retina, excite a) the pretectum, which exerts a feedback control on the pupil diameter, and b) the hypothalamus, which acts on the circadian clocks.

The Alert Pathway: 10% of the sustained ganglion cells and all the brisk ones, after a relay within the LGN, reach the superior colliculus without passing through the cortex (Fig. **3.5**). A rough analysis pathway starts from the colliculus and goes through the pulvinar (a thalamic nucleus) to the amygdalae [9] and to the cortex.

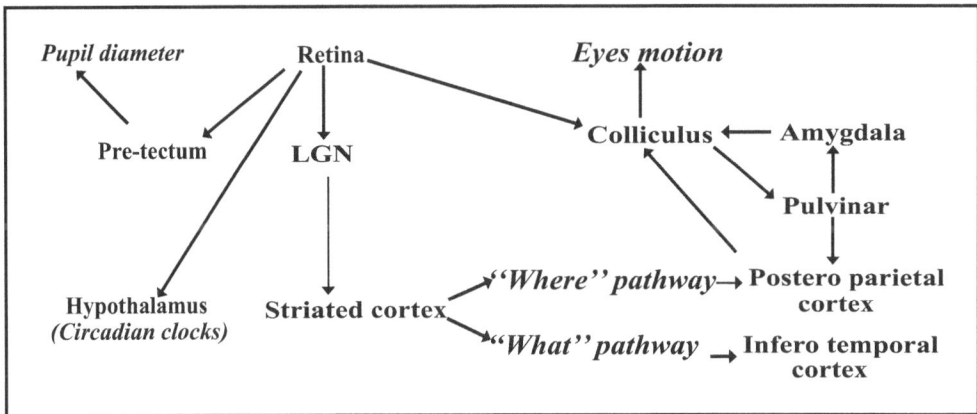

Fig. (3.5). The visual pathways.

The Main Pathway: All the axons reach the lateral geniculate nucleus (LGN) of the thalamus, the striated area 17 of the cerebral cortex (Figs. **3.5**, **3.6**). Areas 18 and 19 are excited by the area 17. (In cat, LGN has direct connections with areas 18 and 19). The striated area 17 receives through the LGN signals from ganglions S and B and transmits them to the peri-striated areas 18 and 19 (Fig. **3.6**) which are devoted to the processing and the recognition of visual signals. Their stimulation can induce hallucinations, their lesions various troubles such as distorted perceptions or many types of agnosia (loss of some recognizing ability).

The course of signals through these areas makes an arborescence of parallel pathways. A dorsal pathway is in charge of the localization of objects and of the analysis of their motions. Area 7 links peripheral vision and extra corporal exploration. It is connected with the thalamic pathway. Area 8, strongly coupled with the superior colliculus, governs the motion of the eyes through the ocular muscles. A ventral pathway is in charge of fine recognitions. It needs time, so it cannot use brisk signals. Shape and color are separately processed in areas 18 and 19. Some bilateral lesion induces a visual aperceptive agnosia (impossibility to name an object). Another lesion of left hemisphere leads to color agnosia (shape is recognized, but the object seems grey). Area 39 in the left hemisphere is partly devoted to reading. Inferior temporal cortex of the right hemisphere is partly devoted to the identification of human faces.

Thus, visual processing seems to be distributed all along the brain.

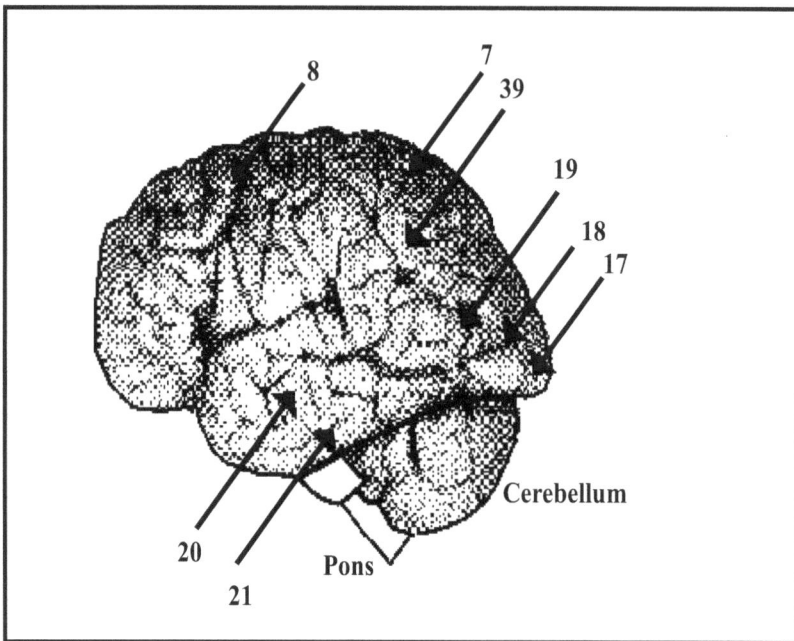

Fig. (3.6). Some cortical areas.

3.1.3.1.4. Functions of the Successive Areas

When following the visual pathway far away from the retina, one finds more and more complex characteristics for the luminous stimuli inducing the response of the various retina maps [10]. For ganglion cells, the most efficient stimulus is a little luminous circle. In the so-called "simple cells" of the cortical area 17, the stimulus is a bar with a fixed orientation. (We improperly call simple, complex, delayed… cells some standard neurons, the outputs of which are simple, complex, delayed… messages).

While receptive fields of area 17 are devoted to one or the other eye, many cells of area 18 are "binocular depth cells": their response is maxima if the fields of view of the two eyes are superimposed.

In areas 18 and 19, one finds "complex" and "hyper-complex" cells, which are sensitive to the length of a bar or to a part of curve, to their velocity and to the direction of their shift. At this level, color signals seem to be separated from the shape ones [10].

At the end of the process, specific neurons are sensitive to the sight of some particular objects. Although it is a difficult task to identify them, some specific neurons have been located: a neuron specific of the face of its keeper has been identified in the inferior temporal cortex of a monkey and neurons sensitive to the number of objects in any collection (but only for small numbers) have been located in primate prefrontal cortex [12]. Each of the recognizing devices is supported by a different cortical localization. Thus, different lesions impair the ability to recognize the identity or gender of a person [13] or the emotion signaled by his face [14].

3.1.3.2. Analysis of Central Vision

3.1.3.2.1. General Principles

Are these facts sufficient to imagine a device working as the real visual device? The key of the discussion is the limited size of the cortical sub-units. It is often assumed that ganglion cells outputs are combined to excite cells responding to colored simple lines, that the outputs of these cells are combined to excite cells responding to "complex" lines regardless of color, and, on the other hand, that the contrast between the outputs of the simple cells is used to excite cells with extremely high color selectivity. But, in this naïve concept, the number of combinations of N inputs would be much greater than N and would exponentially increase from a sub-unit to the following one. The device needs a drastic economy of the number of outputs in each sub-unit.

The economy is obtained by fragmentation of the processing in both space and time, by separation of circumstantial parameters (such as the direction and distance of the examined object, the head position, illumination) from specific signals invariant under a change of the circumstantial parameters such as the shape of the object, and by the intensive use of competitive inhibitions and choices (see paragraph 3.3.4). More and more detailed data are successively gathered by a scanning process. Then, a learned device (similar to the one used to recognize smells) allows recognition.

3.1.3.2.2. Searching for the Vertical

The first sub-units of the area 17 carry out a first adjustment of the control parameters. The inputs are the signals of ganglion cells. Each one is associated with the coordinates (x, y) of the center of its receptive field, with the shape and the colors of this field and to an illumination range. Linear combinations of their signals excite intermediate neurons (the columns of the sub-units), each of them responding to the illumination of some half plane (simple contrast) or of some bar (double contrast). These bar supporting neurons are characterized by their center (x',y'), the shape and colors of their fields, by their illumination range and by a new parameter, the orientation θ of the bar.

In most of cases, several bars are simultaneously excited. A competitive modulation chooses the more numerous and bright almost vertical bars (angle θ_V). (If there are no visible vertical lines, the system chooses almost horizontal ones). This angle θ_V is a rough indication of the angle between the head axis and the vertical. It is exported as a circumstantial parameter and participates to the control of the head motion. On the other hand, it is used in a conditional connective set (Appendix E) to build vertical corrected outputs: this new set contains as many neurons as the input set (made of the ganglion cells). Each input neuron characterized by its center (x, y), the shape and colors of its field and its intensity range is associated with an output neuron, same characteristics except the coordinates (x_O, y_O) of its center. They are the position that would have the luminous spot on the retina if the head was upright. (x_O, y_O) is deduced from (x, y) by a rotation of angle θ_V. The output signals are invariant under a change of the head inclination.

3.1.3.2.3. The Physiological Zoom

Recognition of objects cannot and does not depend on their distance (of their apparent size). The brain ability to change the apparent size is sometimes called "physiological zoom". It works almost as the electronic zoom of a digital camera. The colliculus selects a window (size L_0, center x, y) on the retina. Most of times, L_0 is inferior to the size before zooming L_{MAX}. Signals coming from points outside the window are not transmitted. Too detailed signals inside the window are not transmitted now, but will be later studied during scanning (Appendix F). We discuss hereafter only the central vision. Its size is 4 d°, its angular resolution

1'. Thus, it contains $240*240 \approx 6.10^5$ pixels (each pixel containing about 100 neurons coding for illuminations and for color and shape of the receptive fields with the same center x, y). The output signals are associated with wider receptive fields, so they are less numerous (about 100 signals, each supported by 100 neurons coding for illumination, color, shape). Such a weak number makes easier the combinatorial analysis needed to recognize squares, ovals and so on.

3.1.3.2.4. Scanning

During scanning, the size L of the window can only decrease, but the initially chosen value L_0 is kept in a short-term memory: the eyes search for sharp details within the target, for instance for a little red square in a big green one. The eyes scrutinize now on the little one. A very little saccade brings to the center of the retina a point which was before (x_0, y_0) and the L value, which was L_0 becomes L_1 $< L_0$, L1 matching the size of the little red square. The ratios L_1/L_0, x_0/L_0 and y_0/L_0 (size and position of the little square relative to the big square) are invariant parameters, which can be used as possible components of a signature. They are kept in short term memories.

Contours, colored masses, black and white shapes are processed in several parallel pathways under a condition: at the same time, all of them use the same zoom choices.

3.1.3.2.5. Group Recognition

From this point, the device uses a system of words and passwords similar to the one described for the sense of smell. Passwords (called also "signatures") are stored as a list of shapes in inborn libraries. Very rough inborn passwords (an oval for a human face) are used to direct the signals towards specialized recognizing areas: recognition of objects, of human faces and reading are in charge of three different cortical areas.

3.1.3.2.6. Detailed Recognition

Detailed recognition process uses either inborn passwords (the recognition of human moods from analysis of the face is universal, hence we know it is inborn) or more and more detailed learned passwords (we learn progressively to differentiate human faces). Note that, in many cases, one or the other of several signatures of the same object (contours, colored spots) carried by several parallel pathways is sufficient to trigger the recognition.

3.1.3.2.7. An Example of Processing

The next paragraph will be devoted to the visual choice. For now, we assume that the eyes direction and the L values are well chosen. Hyper-complex cells detect rough contours. An oval is taken as a possible signature for a human face. Then, the retina map is transmitted to the inferior temporal cortex, which proceeds to 3 simultaneous operations: a) Analysis of the curvatures of the eyebrows and of the mouth characterizes the mood of the looked at human. The meaning of the different grimaces is universal: we think it is an inborn way to communicate similar to the use of pheromones by many animals. b) Sex recognition. c) By using a learned kit of simple shapes recognition, identification of the person. Note that the value of the first zooming ratio L_0/L_{MAX} can be used as a learned evaluation of the object distance.

3.1.3.3. The Mechanisms of Visual Choices

Choosing is a very general phenomenon. Section 3.2 will be devoted to its study. In this paragraph, we just describe the observed facts for vision. The superior colliculus is in charge of the visual choices. Several quantitative signals (the visual drives) are the inputs of a competitively inhibiting device, which chooses the most intensive input. The (quantitative or all-or-none) output associated to the chosen input is the only to fire. It induces a) a new choice of L, x, y, and b) some motions of the head and the eyes.

3.1.3.3.1. Response to an Alert Signal

The main competitors are the quantitative signals coming from the main pathway and those coming from the alert pathway: the signals of brisk ganglion cells are directly transmitted to the colliculus by high conduction velocity axons. The only information transmitted is the direction (x, y) of an unidentified object. The intensity of this drive is modulated by fear and anxiety centers (which are in the deepest layer). If the drive associated with an alert in the (x, y) direction is chosen: a) A large amplitude saccade turns the eyes towards the expected direction of the unidentified object. b) The alert drive is cut out. c) A scanning order is triggered; the short term memories supporting old values of the various parameters are ribbed out; defect values (eyes looking to the infinite, L_{MAX}) are chosen; the search for the object begins.

3.1.3.3.2. Reinforcement of Alert by the Thalamic Analysis

While the cortical detailed analysis of human faces leads to the identification of the observed man and to the detection of his mood, the pulvinar analysis of the alert pathway signals (Fig. **5.5**) leads to recognizing human faces and to detecting possible fearing or threatening moods. It is a rough analysis, using only low spatial frequencies. The signature of the mood seems to be linked with the appearance of the eyes [15]. When such an alarming mood is detected, the pulvinar excites the amygdalae which reinforce the alert drive in the choosing colliculus.

3.1.3.3.3. Locating the Target

The most intense signal (associated with the biggest, the brightest or the most mobile spot in the wide explored visual field) is chosen. A little saccade brings precisely the eyes in the good direction. The eyes accommodate and their convergence is optimized (see 3.1.3.4 below). The parameter L_0 of this choice is taken as initial value for the scanning. It is kept in a short time memory. Scanning starts: the eyes search for more and more sharp details within the target.

3.1.3.3.4. Going on with Scanning

Sometimes, we do not recognize the target. Disappointment (see III-2) cuts out the drive or causes a return to the L_0 window and a second scanning explore some new details in a slightly different direction. For instance, shape of the nose and shape of the mouth are successively analyzed. As they are kept in memory, they are used simultaneously for recognition.

If the target is recognized, the exploring drive is cut out. But, if it is found "interesting" by some frontal area, a feedback signal is used as a new drive and the observation of the target goes on (for instance, when we read a thrilling novel; then, alert signals do not disturb us). In other cases, recognition cut out the scanning (finding the name we were searching is a satisfactory signal).

3.1.3.4. The Control of the Visual Apparatus

While the control of pupil diameter is an easy problem, the accurate control of the direction and of the curvature of the crystalline lens of both eyes requires some attention.

3.1.3.4.1. The Physiological Nystagmus

The fine adjustment of the eyes begins when the target has been located: a central spot has been chosen. Its contour excites for each eye separately some simple cells in area 17 and the contour contrasts are amplified in area 18. Physiological nystagmus is a constant, weak and fast shaking of the eyes direction, generating small variations of these contrasts. Their variations allow optimizing the parameters values by acting through the motor centers on the various eye muscles.

3.1.3.4.2. Separated Adjustment for Each Eye Focal Length

Blurring (due to an imperfect focusing) decreases the contrast of the image received by the retina and therefore the intensity of the signals in area 18. The curvature of the crystalline lens is adjusted for the maximum intensity.

3.1.3.4.3. Separate Adjustment for Each Eye Direction

Once the object has been chosen, the best eyes direction is such that the center of the contour is at the center of the retina. Distance to this optimum is measured by area 18 and corrected. This loop governs a) the convergence of both eyes (furnishing an evaluation of the distance between the object and the observer) b) the pursuit of a slowly moving target.

3.1.3.4.4. Parallax Correction

In area 17, cells devoted to right and left eyes are very close to prepare a mixing of their signals. This mixing requires that adjacent Right and Left cells are associated with the same point of the target. A loop associating the LGN and the area 17 allows a satisfactory mixing (Appendix G). Note that, after parallax correction, a local disagreement between x_{LEFT} and x_{RIGHT} generates a 3 dimensional sensation.

3.1.3.5. About the Detection of Motion

3.1.3.5.1. Predictive Function of the Dorsal Visual Pathway

A dorsal cortex pathway is mainly devoted to the visual analysis of position and motion in order to predict the trajectory of a target seen by the foveal retina. Another goal of the system is to detect the movements of the head and of the body. The output signals are mixed with the vestibular signals and used to maintain the equilibrium.

The device uses the signals of both brisk and sustained ganglion cells. Messages from a brisk cell mean: a contrasted spot is moving in (x, y), but does not furnish any information about the magnitude and the direction of the velocity vector. The processing of the visual signal is very similar to the processing of radar plots: sequences of plots let to compute the trajectory of one punctual target, to predict its position in the next future and to correct preventively the weapon and antenna adjustments. In the simple case of eyes looking at a punctual target, the direction is measured at two successive times: (x_1, y_1) at time t_1, (x_2, y_2) at time $t_2 = t_1 + \delta t$. Predicted position (x_{3PRED}, y_{3PRED}) at time $t_3 = t_2 + \delta t$ is computed by using the detectors of sequential signals described in 1_3_4_5. Then, a saccade brings the eyes in the predicted direction before the time t_3.

The alert pathway uses a similar device to detect very bright and fast moving targets.

3.1.3.5.2. The Multi Targets Problem

Tracking multiple targets with radars or other type of sensors is a difficult problem to solve for system engineers. They have elaborated complex statistical algorithms or some heuristics to obtain acceptable solutions. But none of these solutions are able to give the right answer with a 100% probability. The main problem is to always associate an unidentified detection within a swarm of targets to its proper track.

3.1.3.5.3. Partial Visual Detection of Our Body Motions

In man, the problem is solved by associating the signals of body sensors to the visual signals to characterize the motion: turning the head is different of bending it, and each of the possible motions is fully characterized by the value of only one parameter (for instance, the rotation angle of the head); thus, the device has only to evaluate this parameter.

3.1.3.5.4. Detecting the Motion of an Object

A similar device is used although, in this case, the brain does not own auxiliary sensors to precise the $1 \rightarrow 2$ connection law. It uses hypothesis or learned data. Rigidity hypothesis seems to be inborn. Some stereotyped deformation laws seem to be learned. Note that long term prediction of trajectory seems a learned extension of the inborn short term one: babies learn to follow (and to predict the position of) an object passing behind a screen.

3.1.4. Some Words About Hearing

Main results: The principles of hearing and visual analysis are very similar.

3.1.4.1. Hearing Usefulness

The function of the hearing system is to extract some significant neuronal signals from air vibrations to trigger some behavior. More precisely:

a) Hearing allows social communication: the dog recognizes various type of barking, the man various phonemes.

b) It allows to locate and recognize either strong alarming sounds or weak intensity noises when hunting or hunted.

3.1.4.1.1. Hearing is Not Used in the Same Way by Man and Less Evolved Mammals

From the observations of deer hunters [16], deer use mainly hearing to be alerted by weak unusual noises. Great adjustable ears allow strong sensitivity and accurate direction measurement. Learning of usual noises allows inhibiting the alert pathway when unnecessary. On the contrary, alert is only a secondary function of human hearing, the main function being the thorough analysis of usual noises. With its unmoving ears, man owns a poor sensitivity to weak noises and poor direction measurements, but proceeds to an accurate spectral analysis. Dog hearing seems at midway between man and deer: its alert sense is as acute as the alert sense of the deer, but it gives some attention to usual noises.

3.1.4.2. The Ears

3.1.4.2.1. The Middle Ear as an Impedance Adapter

Incoming sounds are atmospheric pressure waves. Inner ear is sensitive to pressure waves in a physiological liquid. The middle ear enables a transmission from a medium to the other with weak intensity losses. Muscles of the middle ear govern the device rigidity. They are imbedded in a protective loop, the function of which is similar to the function of the eye pupil: in case of too strong an acoustic intensity, they act to decrease the transmission towards the inner ear.

3.1.4.2.2. The Inner Ear

The inner ear contains the vestibular apparatus (a 3D accelerometer, which governs the sense of equilibrium; we will not describe it) and the **cochlea** devoted to hearing, which contains the Corti organ made of a rigid membrane and a vibrating one (the **basilar** membrane) linked by sensorial hair cells.

The Basilar Membrane: It is an inhomogeneous plate. When excited by an oscillating pressure with a sound frequency f, the amplitude of the induced vibration of the plate is negligible except around some location x of the plate (where x is measured from the apex of the plate towards its narrow basis) and x is almost proportional to Log (f). Thus, the plate carries out a rough frequency analysis of the incoming sound [17].

The Inner Hair Cells: They are two types of hair cells: the inner and the outer ones. Inner hair cells are the transducers. They are mechanically linked to the basilar membrane: the hairs of each of them are moved by the local plate vibrations, opening and closing ionic channels, thus generating a cell potential depending on the local vibrations of the basilar membrane. So they are sensitive to a narrow band of sound frequencies.

The hair cells are hypersensitive: They are three compartments systems: the cell body, the usual external medium (the perilymph) and an unusual external medium, the endolymph. The motor of hypersensitivity is an active Na^+/K^+ pumping from one external medium to the other (Appendix H). The hypersensitivity confers on them two properties: a) they act as rectifying diodes (the mean ionic signal is not zero although the mean pressure signal is zero). The inner hair cells release transmitter only when the hairs are elongated, that means during half the time of the mechanical vibration. For very low frequencies, this fact leads the nerve fibres to give out intermittent burst of spikes. b) They have a very great sensitivity for weak intensity sounds. The hair cells measure Log (I) rather than I.

The Outer Hair Cells: One observes a great number of outer hair cells, which own practically not any nervous output, but receive signals from the brain and contain contractile proteins. They act on the basilar membrane to increase the amplitude of the local vibration in case of a weak intensity sound (behaving as a

30 db automatic gain control) and to enhance the contrast between close frequencies.

3.1.4.2.3. The Auditory Nerve

The **bipolar cells**, the axons of which form the auditory nerve, are transient swift neurons. (Do not confuse these neurons with the hypersensitive bipolar cells of the retina; ear does not own ganglion cells). Their connection with hypersensitive cells requires an intensity coding using (as in the case of the eye) a progressive recruitment.

3.1.4.2.4. Performances

The *frequency range* of hearing is from 20 Hz to 20 kHz in man, from 30 Hz to 45 kHz in cat and from 65 Hz to 40 kHz in dog. (Note that the frequency spectrum of speech ranges from 100 to 200 Hz for man and from 150 to 300 Hz for woman). High frequencies are most of time due to impulse-like sounds: the spectrum of a clash of cymbals ranges from 250 Hz to more than 10 kHz.

The *intensity range* is 120 db for man with acuity of 1 db. The acuity range is probably greater in the dog, in relation with its ability to orientate its ears.

3.1.4.2.5. A Comparison Between Man, Cat and Dog

The man owns 3500 inner hair cells, 20000 outer hair cells, 28000 bipolar cells; the cat 2600 inner cells, 10000 outer cells, 50000 bipolar cells; the dog only 1600 inner hair cells, 6100 outer cells. From the above numbers, we remark that the dog (which has a very efficient hearing) owns a very weak number of sensors. To discuss this surprising fact, we compare (in Table **3.3**) for our three examples the number of octaves (NbOct) in the hearing range, the ratio of the number of outer to the number of inner hair cells (O/I). Because hearing obeys to the logarithmic Fechner's law, we can deduce the frequency discrimination $\Delta F/F$ from the number of octaves.

Table 3.3. **Hearing performances of men, cats and dogs.**

	NbOct	Outer/Inner	$\Delta F/F$
Man	10	5.7	$2\ 10^{-3}$
Cat	10.6	3.9	$3\ 10^{-3}$
Dog	9.3	3.8	$4\ 10^{-3}$

The frequency ranges for man, cat and dog contain the same number of octaves. The human discrimination is twice better than the dog's one. The number of outer hair cells is especially great in man. We interpret these facts by assuming that external cells enhance the contrast between close frequencies and that the man needs great acuity and great contrast to distinguish phonemes, but that dog has a better sensitivity to weak sounds (and, we are to see it, a better ability to found the direction of the acoustic source).

3.1.4.3. Brain Processing

3.1.4.3.1. The Pathways

Entering the medulla at the inferior border of the pons, the nerve fibres immediately bifurcate in the cochlear nucleus to reach several olive nuclei in the pons. Then several parallel pathways go to the inferior colliculus, the medial geniculate nuclei and the primary auditory cortex (areas 41 and 42), which is adjacent to the Wernicke area (area 22), which governs language understanding.

3.1.4.3.2. Generalities About the Processing

As for visual processing, the hearing processing is a sequence of divergent-convergent transformations made by several parallel pathways. The matrix-like aspect of the circuitry is supported by a systematic topology. Thus, one observes all along the pathways many maps of the basilar membrane. The processing builds up more and more complex signals. For instance, some neurons sensitive to both the direction and the velocity (which are useful for hunting) of a noise have been observed in a dog cortex [18]. Many transformations use delayed neurons allowing sequence recognition (see Chapter 1.3). Delayed neurons are especially numerous in the cerebral cortex [19]. The processing tends to build up invariant signatures: a melody is recognized after an octave transposition, a phoneme whatever is the tone of the voice. Then, by adding several sequence detectors and eliminating useless parameters, the device is able to recognize tunes and sentences.

We will see in 5.1 that the ability to identify phonemes is an inborn property of the circuitry, but that the list of the few recognized phonemes results from learning.

3.1.4.3.3. Locating a Sound Source

Localization process begins from the end of the auditory nerve. Moving the ears to maximize the intensity received by each of them is the most accurate and the most robust locating device. Unluckily, man is devoid of such a device. He owns (as the other mammalians) two devices at the olive nuclei level. They give indications, which have then to be processed. The first one uses high frequency sound components (more than some kHz). It measures the intensity ratio between left and right ear. Remark that this ratio is greater when the source is close to the observer. The second one is an interferometer device using low frequencies (less than 300 Hz, see Appendix I). For these frequencies, nervous fibres deliver bursts synchronized with the sound. The device separately measures for each sound frequency the phase difference between right and left signals. (These two devices do not specify a direction, but a cone of possible directions).

To be usable, these results have to be processed. By moving the head, one obtains a sequence of observations (with various possible directions cones). From such a sequence, several transformations allow to extract the direction and the motion of the sound with a better accuracy than after an isolated measurement. In the colliculus of the owl, a map of auditory space has been detected [20]. In man, sound sources are probably reported on a visual map.

3.1.4.3.4. An Alert Pathway

It has been found [21] that, in some part of the inferior colliculus, the neurons are not sensitive to pure sinusoidal sounds, but only to intense and brisk noises. We think that, coupled direction signals, they act on the superior (visual) colliculus to induce an alert behavior.

In summary, the hearing of dog is optimized to deal with hunting, the hearing of man to deal with language.

APPENDICES TO SECTION 3.1

A: From the illuminations to the opening of post-synaptic channels

B: The hypersensitive visual cells

C: The mechanism of brisk ganglion cells

D: Quantitative schematization of the retina interconnections

E: Conditional connective sets

F: The physiological zoom

G: The parallax correction of eyes directions

H: The hypersensitive cells of the ear

I: Phase difference measurement by the human ear

Appendix A: From the Illuminations to the Opening or Closure of Synaptic Channels

The Photonic Flow on a Cone

The eye receives an illumination I (cd/m^2). Without diaphragm, the optical apparatus sends to the retina a flow Φ (photons per square micrometer and per second) proportional to I. For man, $\Phi = 450$ I. In man, but not in honeybee, the closure of the pupil reduces the light input area by a factor a. Then, $\Phi = 450$ a.I. The relative area a (equal to 1 for weak illuminations) is a $= 1$ in honeybee. In man, experimental results are well fitted by:

a $=1$ if I $< \mathbf{2\ 10^{-2}}$, a $= 0.07$ if I $>\mathbf{10^3}$. Between these limits, $\mathbf{a = 0.628 - 0.186\ \log_{10}\ I}$

The Number of Activated Opsines

The N_o opsine proteins of a photoreceptor have three states: a) a usable state (N_U molecules); b) when excited by a luminous flux Φ, the absorption of light (cross section σ) leads first to an activated state (N_A molecules), which has a very weak decay time τ; c) an inactivated state (N_{IN} molecules). A slow metabolic process (time constant T) makes it to return to the usable state.

The system obeys to the equations:

$$\mathbf{dN_U/dt = -N_U\ \sigma\ \Phi + N_{IN}/T \quad dN_A/dt = N_U\ \sigma\ \Phi - N_A/\tau \quad N_U + N_A + N_{IN} = N_0}$$

After a steady illumination Φ, the concentration of usable proteins decreases toward the value: $\mathbf{N_U = N_0/[1 + \Phi/\Psi]}$ with $\mathbf{\Psi = 1/\sigma T}$ and the concentration of activated opsines is:

$N_A = \Phi.\tau\sigma N_U = \tau\sigma N_0\ \Phi/[1 + \Phi/\Psi] = \tau\sigma N_0\ aI/[1 + I/J]$ with $J = 1/450\ a\sigma T$

Appendix B: The Hypersensitive Visual Cells

Qualitative Description

The cell potential V depends on the balance between the Na expelling by active pumping and. the Na entrance by passive channels, which is proportional to the ratio Ω of opened channels. The special properties of these cells result from a non linear phenomena: the binding of activated opsine molecules with v open Na channels ($v = 2$).

A Simple Model:

Let N_C be the total number of Na channels in the cell. Then:

$$\Omega^v\ N_A/N_C = k(1-\Omega)^v$$

where k is equilibrium constant and where N_A was evaluated in the appendix A as a function of the illumination and of the parameter $J = 1/450\ a\sigma T$). On the other hand, we assume that the leakage obey to a linear Ohm's law with a Nernst potential $V_{Na} = +40\ mV$ and that the pump obeys to the simplest law, an all-or-none one: its current is zero if $V < -70\ mV$ and 1 if $V > -70\ mV$. Then, the electric balance is $\alpha(40-V)\Omega = 1$ where the parameter α depends on the ratio of pumps and channels numbers. It determines the potential in darkness: $\alpha(40-V_{Dark}) = 1$. This system of equation allows to easily compute I as a function of V. It shows that $10J = I$ at saturation.

Results:

We know that the cell potential V is -40 mV in darkness for both rods and cones. The best fit is obtained with $v = 2.1$ (probably a rough evaluation for $v=2$). The computation gives for rods and for cones (Table **3.4**):

Table 3.4. Current *versus* potential in hypersensitive visual cells.

V	-40.07	-40.15	-40.3	-41	-43	-50	-60	-70
I rod		$6.7\ 10^{-4}$	$2.8\ 10^{-3}$	$3.6\ 10^{-2}$	0.36	5.0	32	500
I cone	$6.7\ 10^{-3}$	$3.3\ 10^{-2}$	0.14	1.8	18	$2.5\ 10^2$	$1.6\ 10^3$	$2.5\ 10^4$

Appendix C: The Mechanism of Brisk Ganglion Cells

We assume that the [+] and the [-] receptive fields of an ON brisk ganglion cell have the same location and that there is not any threshold modulation. The firing results from an instantaneous excitation and from a slow inhibition (Fig. **3.7**).

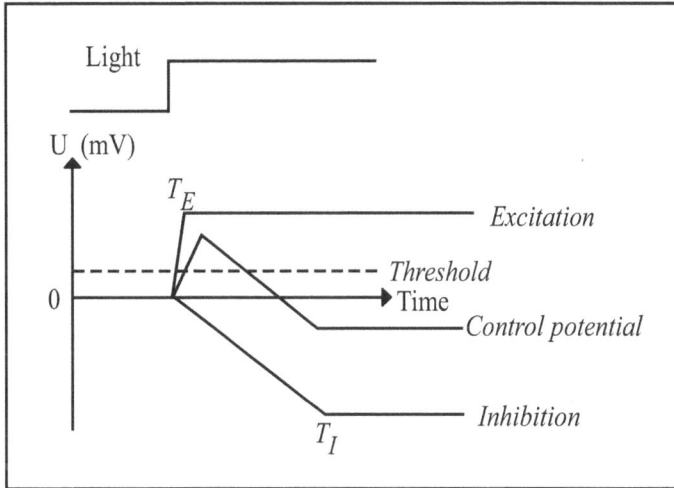

Fig. (3.7). Control potential in a brisk ganglion cell.

In static conditions, the inhibition is stronger that the excitation, the cell is mute. The synaptic delay T_E of the exciting bipolar cells is weaker that the synaptic delay T_I of the inhibiting bipolar cells, and both are weaker than the time constant T_D of the ganglion cell transient adaptation. If suddenly enlightened, the neuron control potential is great between times T_E and T_I and the response $F_0(U)$ is very great. After T_I, U is negative and the cell is mute. Symmetric assumptions apply to the OFF brisk cell.

Appendix D: Quantitative Schematization of the Retina Interconnections

The Interconnecting Matrix

We have to describe the circuitry between N cones and the same number of nerve fibres with $N \approx 10^6$. The N inputs are the light intensities stimulating the cones, the N outputs the spikes frequencies of the fibre. It is more handy to look at relative numbers I = intensity/intensity max and F = firing rate/firing rate max. The ranges of I and F are from 0 to 1.

Between the inputs and the outputs, we would have to describe a lot of transformations (control potential of the cones versus the received intensity, *synaptic excitation of the exciting and of the inhibiting bipolar cells and so on). We just remark that these relations are roughly linear. Then, the connecting circuitry can be described by a set of positive and negative weights and the input/output relation by a matrix: Outputs = Matrix*Inputs (or O=M.I). But the real output F has to be in the possible range 0,1. So, the usable relation will be:*

F=0 if M*I < 0; F= M*I if 0 < M*I < 1; F=1 if M*I > 1

We have now to write a realistic M matrix.

Block Matrices

Each cone is labeled by its coordinates x, y on the retina (x takes $\sqrt{N}$ values) and each fibre by its associated coordinates X, Y. In the more naïve trial, $M_{xyXY} = \delta_{xX} \delta_{yY}$ which means $F_{XY} = I_{x=X, y=Y}$. (We use here the Kronecker's index δ_{AB} which is 1 if A=B and 0 if A≠B) From now, we look at block matrices: n cones project their data on n ganglion cells. The total matrix is made of N/n blocks.

First Step: Introducing the Intensity Coding

The output of one cone has to be coded by progressive recruitment of 6 fibres. (Then, from a GC to the next one, the light intensity corresponding to the threshold of the fibres is multiplied by 10). As the number of fibres is equal to the number of cones, 6 cones must excite one fibres by adding their signals. Thus n=6. *Inside a block, the cones are labeled by the index q (q=1 to 6), the GC by an index Q. Each block of cones is labeled by the indexes x, y, which are now the coordinates of the center of the block (x takes now $\sqrt{N}/6$ values). Then, the matrix component becomes:*

$M_{xyqXYQ} = \delta_{xX} \delta_{yY} k^{1-Q}$ *with k = 4 and the output O=M* I is*

$$O_{XYQ} = \sum_{Q=1}^{Q=6} 10^{1-Q} I_{x-X, y-Y}$$

Second Step: Introducing Inhibitions and Balanced Fields

We search a linear transformation that shall satisfy the following requirements:

-        It is invertible in order not to lose information

- A maximum number of the output components are null under a diffuse light

- Symmetries between the vertical and the horizontal directions are preserved as much as possible

To do so we part the set of cones in related subsets of 4 fields of 6 cones (6 cones of one center field and 18 cones of three peripheral fields). We do the same for the set of fibres that we part in related subsets built on the 4 groups of 6 fibres related to each specific field of cones. On the product of a subset of cones and a corresponding subset of fibres, we can build a 6 by 6 M matrix relating the 6 recruited fibres to a vector with 6 dimensions and identical components. For each of the 4 subset of fibres, four of these matrices can be built and in total, a 4 by 4 block matrix, using M as a block, relates an output vector of 24 fibres to an input vector of dimension 24 with four groups of 6 identical components.

The weight of the blocks on the same line must be such that the requirements that have been expressed here above are satisfied and that the maximum intensity received by one group of cells does not exceed its possible range: 1. This impose the 4 by 4 matrix of Table **3.5**.

Table 3.5. A canonical coding matrix.

0.25	0.25	0.25	0.25
0.5	0.5	-0.5	-0.5
0.5	-0.5	0.5	-0.5
1	-0.5	-0.5	0

The first line respects symmetry between x and y, but is not balanced: it is sensitive to the total luminous intensities. The three other lines give a null output in diffuse light. The last one is the schematization of a round ON field and a round peripheral OFF field. The lines 2 and 3 schematize vertical or horizontal elongated fields.

The block matrix is invertible since, due to the specific structures of the blocks, the column vectors of dimension 24, as well as the line vectors, are clearly linearly independent.

Step 3: Introducing Colors

The above matrix describes a black and white circuitry. We take into account the three colors of the cones pigments. The balanced round fields have an inside field associated to one color, a peripheral field associated to another (6 combinations). On the other hand, unbalanced fields and elongated balanced fields work in black and white by adding signals from the three types of cones.

To treat these 9 cases, we look now at little blocks (internal index q) of 6 cones of the same color c (c = 1 to 3) and to great blocks containing 3 little blocks (index p = 1 to 3) of each color. Little blocks of ganglion cells (index Q) contain 6 progressively recruited fibres. Great blocks (index P = 1 to 9) contain 9 little blocks of fibres. x takes now √N/54 values. In the new μ matrix, the fourth line of the above matrix is replaced by 6 lines: two lines for each value of c, one of them (p = 1) with negative coefficients corresponding to the color c+1, the other (p=2) to the color c+2. The three other lines are formed by summing the signals of all colors.

Step 4: Introducing the Symmetry Between [+] Field and [-] Field Inside

We introduce a last parameter ε (ε = +1 or -1). The matrix components for ε = +1 are the same as above, for ε = -1, they are multiplied by -1. x takes now √N/108 values.

Unbalanced, C and D are balanced and unsymmetrical. Thus, we obtain N/4 balanced symmetric signals, N/2 balanced unsymmetrical ones and N/4 symmetric unbalanced.

Appendix E: Conditional Connective Sets

The inputs are N_1 neurons, each of them identified by a set of labeling parameters (for instance x, y, colors of the receptive fields and so on). Linear combinations of the input signals excite a set of $N_1 N_2$ intermediate neurons identified by the same set of labeling parameters and by a new parameter (for instance the bar orientation θ).

The sum of the signals of all the N_1 intermediate signals with the same value θ acts as the weight w(θ) of this parameter. The greatest weight is chosen by some competitive inhibition, defining a choosen value Θ. Now, all the intermediate neurons with the same (x, y, colors and so on) parameters are connected to N_1 output neurons. But, a competitive inhibition (see III.3.4) inhibits all the links except the ones associated to the value Θ. Thus, the choosen value of the new parameter is transmitted by a lateral pathway while the firing of an output neuron means that a bar with the choosen, but here unknown orientation, excites the retina at the place x, y.

Appendix F: The Physiological Zoom

The signals of several original pixels are added to obtain wider size pixels. Zooming without distortion implies that the sizes increase as a geometrical series. Let B be its basis and Λ the size of the retinal field. The size of a pixel is $L_k = \Lambda/B^k$ (where k is an integer and k = K for the original pixels). A window is defined by a number k. It is divided into B^{2n} little squares. Without sufficiently precise experimental results, we assume hereafter the value B = 2 and n = 3. Then, the sizes of the possible windows are Λ, $\Lambda/2$, $\Lambda/4$...$\Lambda/254$. Each window is divided into 64 little squares. To cover the central retina with windows of size k, one needs 2^{18-2k} windows. Thus, the number of output pixels is $(254)^2 [1+1/4+1/16+...] \approx 1.3 (254)^2$ with k varying from 3 to 9.

Appendix G: The Parallax Correction of Eyes Directions

In area 17, cells devoted to right and left eyes are very close to prepare a mixing of their signals. This mixing requires that adjacent Right and Left cells are associated with the same point of the target. A loop associating the LGN and the area 17 allows a satisfactory mixing *Even if both eyes aim at the center of a nearby object, a lateral point of the object is projected to different points of the two retinas: $x_{LEFT} = x_{RIGHT} + \partial x$. The parallax correction is roughly linear: $\partial x = \eta.x$ where η does not depend on x, but strongly of the distance between the object and the observer. A divergent-convergent device forms first all the possible $(x+\eta x, y)$ values for all the possible values of η. Then after the choice of only one value, a transformed map is built. (We think that this mapping is a function of the LGN). A rough value of η is given by the angular convergence of the eyes.*

Appendix H: The Hypersensitive Cells of the Ear

They are made (Fig. **3.8**) of three compartments: the cell body, the usual external medium (the perilymph) and an unusual external medium, the endolymph. The motor of hypersensitivity is an active Na^+/K^+ pumping from one external medium to the other.

The pump creates strong ionic concentrations differences between the three media (Table **3.6**; unit = 10^{18} ions/cm^3):

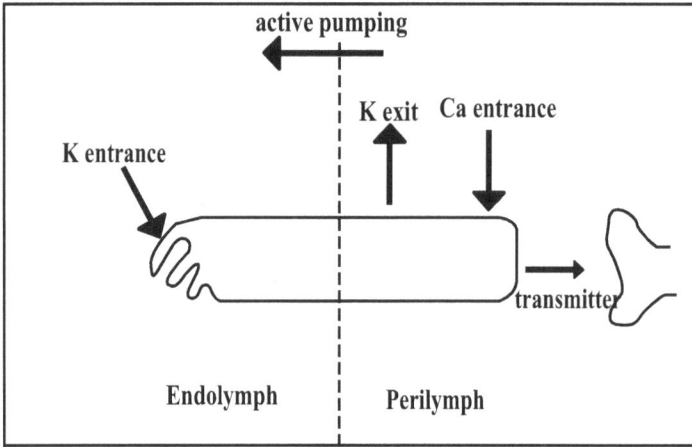

Fig. (3.8). The hypersensitive sensorial hearing cell.

Table 3.6. Ionic concentrations inside the cell.

	Perilymph	Endolymph	Internal Medium
Na^+	86	003.5	13
K^+	03	164.0	74

Potentials at Rest

A great number of Na^+ channels allow Na^+ passive flowing from perilymph to endolymph (Fig. **3.9**).

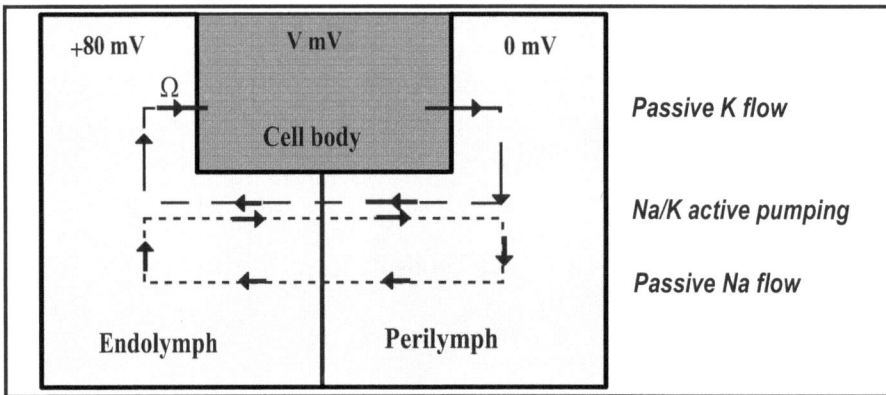

Fig. (3.9). Ionic flows in the hypersensitive cell.

Then, the endolymph potential at rest (80 mV) is close to the endolymph/perilymph Nernst potential. The rest potential of the internal compartment, due to the constant opening of some K^+ channels is -45 mV for a cell body/perilymph Nernst potential: -80 mV. When the hairs are moved by the basilar membrane, their curvature induces the opening of K channels and the entrance of potassium ions. Let Ω_{HAIR} be the ratio of open K^+ channels linking the endolymph to the cell body. A mechanical action (a sound) induces the opening Ω_{HAIR} of a potassium channel linking the endolymph to the internal medium. There is an increase of K^+ ions entrance in the cell. The internal potential changes in such a way that the ionic current outgoing toward the perilymph balances the ingoing one. This change induces an entrance of Ca^{2+} ions in the presynaptic ending and the release of transmitter.

Using linearized Ohm's laws, the balance of potassium currents in the cell body:

$j_{in} = j_{out}$ obeys to: (V - 80). $\Omega = k*(V + 80)$. Fitting k to obtain a great sensitivity to weak sounds, one finds the following example (Table **3.7**):

Table 3.7. Channel opening *versus* potential in the cell body.

Ω	0	0.25	0.5	0.75	1
V (mV)	-80	-26	0	16	27

Appendix I: Phase Difference Measurement by the Human Ear

The distance d between the ears is roughly 15 cm. In air, the sound velocity c = 340 m/s. Thus, the maximum delay between the arrival of sound on each ear is δt = d/c = 4.4 10^{-4} s. The nervous fibre emits high frequency spikes within bursts of duration T/2 separated by silent period with the same duration. The device measures the duration of the overlapping period of the right and the left signals. The time constant of any synaptic device cannot be smaller than 1 ms. So, the overlapping duration $\dfrac{T}{2} - \delta t > 10^{-3} s$ Therefore the minimum frequency of the sound must be some 350 Hz.

A good direction finding cannot be achieved from hearing low frequency sound.

3.2. Instincts, Drives and Choices

Main results: The complexity of instinctive drives. Role of pleasure.

3.2.1. Behaviors and Instincts

3.2.1.1. Behaviors

3.2.1.1.1. Sequential Motor Software

We have seen that the frog retina contains neurons specialized for vultures, flies and worms. The sight of a fly triggers a stereotyped behavior: the frog turns its head, catches, swallows and wipes its mouth. This sequence is carried out till its end even if the prey is missed. In the same way, when a man recognizes a glass of water (or of wine), he drinks it. In both examples, the recognition triggers a stereotyped sequence of stereotypes simple motions (to take the glass, to bring it to the mouth and so on). Each of these motions is oriented by the circumstantial parameters (size, direction of the object). One usually calls **behaviors** the observed response of animals to stimuli.

Now, from the above observation of frog and man, we conclude that the sequence of motor orders is governed by more or less adaptable software. It would be easy for an engineer to build such an automaton by using either neurons or electronic chips.

3.2.1.1.2. Choosing is Necessary

In most of cases, there is a set of possible but incompatible behaviors. The observed one results from a choice. We have to choose when and how to give way to our sexual needs. But we have also to choose between walking or running, between fighting, staying motionless (and almost invisible) or running away, between working or stopping to drink or piss. We choose our words, the direction of our eyes.

3.2.1.1.3. Choosing Boxes

As described in 1.3.4.4, they are neuronal units (such as superior colliculus) with N outputs associated with N inputs. At any moment, only one output is firing, the one associated with the more strongly excited input. The firing output triggers the associated behavior. We call instinctive drive the signal exciting an input.

Note that there is not a sole choice center: the selecting sub-units making choices are distributed within the brain. For instance, the choice between fear and sexual attraction is made in the hippocampus; the choice between looking right or left is supported by the superior colliculus.

3.2.1.1.4. Generalized instincts

We call **generalized instinct** the set of a drive and some sequential software. (Note that learned behaviors are observed: so, we have to deal with learned drives and learned software).

For each generalized instinct, we would have to discuss a) the laws governing the intensity of the drive and b) the nature of the associated software (which can be a sequence of motor orders as in walk or a feedback modulating signal as in visual processing).

3.2.1.2. Connection with Classical Definitions

3.2.1.2.1. Usual and Generalized Instincts

Great instincts (sex, feeding, fear…) are the main motor of animal activity. Our definition is not far from the usual one. But:

a)   Usual instincts are inborn. Generalized instincts can be inborn or learned or more often partially learned.

b)   Many great instincts are fragmented into a sequence of almost independent partial instincts, each of them with an autonomous drive. As examples, look at the nuptial parade of some birds, at the play of a cat with a mouse.

c)   There is a lot of more modest "little instincts", governed by the same scheme, which choose the direction of our look, the next word in a sentence.

d)   To avoid any confusion, we will speak of usual or of generalized instincts.

3.2.1.2.2. Freudian and Generalized Drives

Freud has discovered the important notion of drive. But, for him, it is something like an energy when, for us, it is the firing frequency of some neuron. To avoid any discussion, we will speak of Freudian or of generalized drives.

3.2.1.2.3. A Democratic Analogy

The choices are selected in a sort of democratic way: K. Lorenz [22] spoke of "the great parliament of instincts". In place of using a priori criteria, the brain builds drive signals which are weights depending on the internal state and on the external situation. (Note that they are continually changing).

Sometimes, such a criterion leads to a flat optimum: all the choices lead to the same benefits. Then, the only solution is a random choice and an anarchic organization is as efficient as a democratic one. Look at the displacements of a bank of herrings. In the sea, there is not any privileged direction. Herrings do not own any hierarchy; each of them follows its neighbors. This behavior leads to a random displacement of the whole bank of fishes. If we blind a herring, it becomes unable to see and to follow its neighbors, it swims right ahead, the others follow it, it has become the big chief (this example is not supposed to be a parabola of human political behavior).

3.2.2. Instinctive Drives

3.2.2.1. The Drive Messages are Quantitative Ones

In frog, the fear message is an all-or-not one: when excited, the drive neuron always inhibits the feeding behavior. But, in most of cases, the drives are quantitative signals (transmitted by linear neurons and synapses). This quantitative appearance is often easily observed: the tilt of the ears of a dog is a numerical indication of its mood.

The intensity of a drive is often a simple function of an external parameter. For instance, it can depend on the distance d of some other animal. Thus, in birds, the intensity of aggressiveness creates a repulsive drive which is very strong if d is inferior to some limit d_L and sociability creates an attractive drive decreases weakly with d. When the intensity of the first drive is the strongest, the chosen behavior takes the bird away from another bird; when the intensity of the second drive is the strongest, the chosen behavior brings the two birds closer. The first instinct acts as a short range repulsive force, the second one as a long range attractive force. Their quantitative competition leads to a crystal-like equilibrium configuration with birds regularly placed on a thread with a constant separation d_L.

The same phenomenon can lead to relaxation oscillations. Look at a two years baby. His fear increases with the distance to his mother. But there is some slow synapse in the drive circuitry: the drive intensity needs a delay T_A (3 to 10 s) to increase. When the fear drive is weak, other instincts (curiosity, aggressiveness towards his mother) have strongest drive intensities, the baby goes away. Then, fear increases: after a time T_A, the baby runs towards his mother.

The drives are often the sum of a specific signal and of a more general modulation. Thus, in the visual choice, anxiety favors the choice of any direction in the lateral visual field.

3.2.2.2. Limitations Due to the Finite Range of Excitations

Buridan, a middle age scholar, discussed the case of a donkey situated just between two symmetric bundles of hay. He expected that the donkey, unable to choose, would starve to death. Experiment shows that the donkey eats one, then the other bundle. We observe a failure to choose only when two antagonist drives are so strong that the circuitry is saturated. In a first stage, Pavlov conditioned some dogs to receive a reward when they saw a flat ellipse and to be submitted to noxious stimuli when they saw a circle. In a second stage, he showed to them a large ellipse, perfectly intermediate between circle and flat ellipse. Intense emotional signs were observed, inducing a nervous crisis similar to hysteria for half of the dogs and a hypnotic sleep for the other half.

3.2.2.3. Thirst and Micturition Drives

An engineer would think that drives are only due to the detection of a physiological need. Realty is more sophisticated: drives are made of several components.

3.2.2.3.1. Phenomena Increasing or Deleting the Drive

The human thirst drive results from sophisticated interactions of the signals of several sensors.

3.2.2.3.2. A Specific Endogenous Sensor

Some hormones assume an accurate control of water stocking and excretion. Drinking allows a rough compensation of the water losses of the body and thirst governs drinking: a specific sensor measures the blood osmotic pressure or more exactly its departure from the normal value (the osmotic pressure increases if there is a lack of water or an excess of sodium). The firing frequency of the sensor is proportional to the pressure departure. If it exceeds some threshold, a lateral pathway is excited and thirst feeling becomes conscious. In the case of micturition, the bladder filling (its volume) is detected.

3.2.2.3.3. Psychological Components

The sight (and the recognition) of a bottle and a glass triggers a change in the drive. This change is modulated by the osmo-sensor signal: if this last signal is weak, the sight of the bottle is inefficient; if the osmo-sensor signal is above some threshold, the sight of a bottle triggers a slowly increasing signal. We call it an (generalized) **impatience** signal (Fig. **3.10** below), the pleasure impulse is described lower).

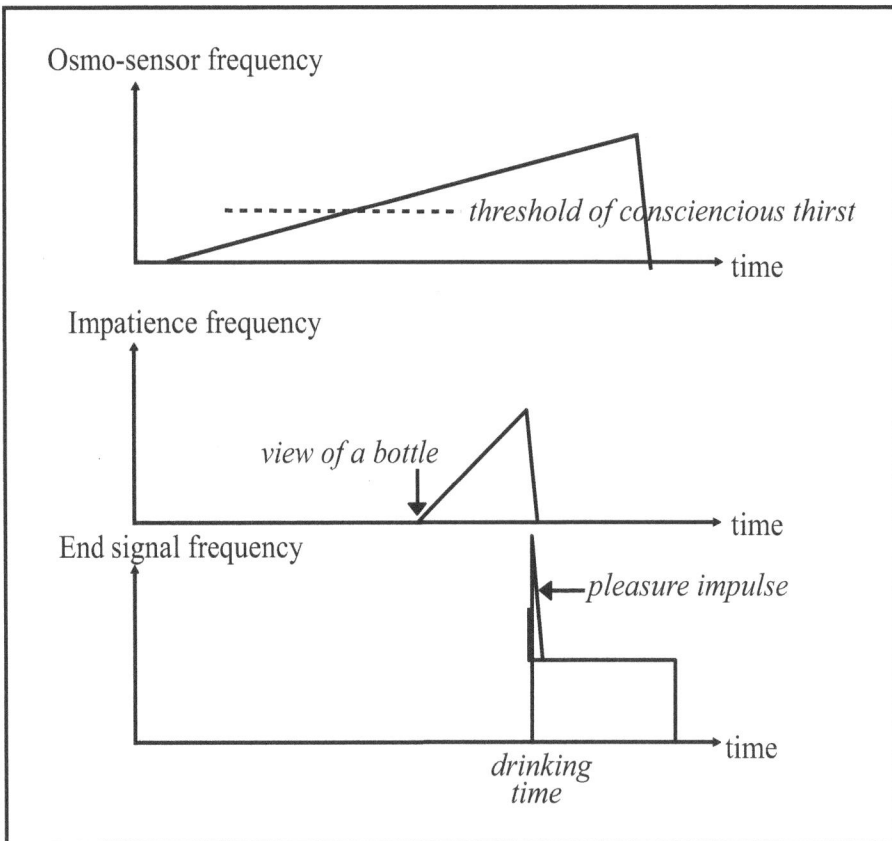

Fig. (3.10). Physiological need and impatience in thirst instinct.

In the case of micturition, some psychological signals are inhibiting: we do not piss on the carpets.

3.2.2.3.4. Deleting the Drive After Drinking is an Active Process

In the more naïve assumption, the drive signal would stop when the physiological need stops. And in fact, the micturition drive stops when the bladder is voided. But this assumption does not work in the case of drinking: there is a 10 minutes delay between drinking and change in the blood osmotic pressure.

The naïve assumption would lead to over drinking. Deleting has to be assumed by a specialized device. A sensor located in the pharynx count a critical number of gulps and then strongly inhibits the impatience and the drive neuron for a time longer than the delay between drinking and change in blood dilution.

3.2.2.4. Other Examples

3.2.2.4.1. End of Feeding Signals

In red fishes, feeding is limited by shortage of food; they do not own an end device. In too rich a medium, they die of over eating.

In mosquitoes, the signal ending the sucking of blood is due to a neuron measuring the abdomen distension. If this neuron is destroyed, the mosquito sucks till it explodes.

In man, the end signal when eating is due to several components: a device similar to the pharynx sensor and a security device measuring the stomach distension. In some cases of obesity (due to a failure of the first system), surgeons put a steel ring around the stomach, thus reducing its capacity. Then, the security device stops the drive for a reduced eaten volume.

3.2.2.4.2. Control of Micturition; Urge Incontinence

The choosing device for voiding is located in the frontal cortex while the behavior is governed by two nuclei in the brain stem, one inhibiting the sphincter during the voiding, the other exciting the detrusor (the bladder muscle). Many women suffer from urge incontinence. One of the symptoms is the need to run to the toilets although they are looking at a thrilling TV movie. The disease is not due to sphincter weakness. Functional magnetic imagery shows that the brain stem works normally, but that the frontal cortex exhibits an abnormal activity. So, in such cases, urge incontinence appears as an excessive weight given to the endogenous drive [23].

3.2.2.5. Special Drive Components

Almost all the simple devices are used as drive components. We give a special attention to impatience, chronometric drives and self-referring regulations.

3.2.2.5.1. Impatience

We call (generalized) impatience a signal increasing slowly after being triggered. Impatiences are used as modulating signals to explore sets of competitive neurons. The increasing firing rate of a first kind of them is due to the progressive accumulation of activator in a slow synapse. Then, they work during some seconds before reaching saturation. Such impatience intervene for instance in colliculus during scanning or in language when we search for a word. A second class of generalized impatience uses long term potentiation, a phenomenon linked to memory (see Section 4.1). Then, the scale of time is about three weeks. Such impatience sometimes leads to obsession.

3.2.2.5.2. Chronometric Drives

Some drives seem not related to any endogenous or exogenous sensor but to be governed by neuronal or possibly hormonal clocks with characteristic times of some minutes or hours. These clocks would be stopped during sleep. They start spontaneously after the ending of the sleep and furnish either a slowly growing signal or a delayed impulse. Such an impulsive signal is a component of the hunger drive (it is time to eat). A slowly increasing impatience explains the need of watching birds in cats and the need for exploring the word in babies, the need of playing with a mouse in cats, of running in babies.

In the same way, end signals or components of end signals are often governed by a clock measuring the time devoted to a particular behavior: in cat, satisfaction is obtained when it has played about ten minutes with a mouse.

Note that random time generators seem to be used sometimes as end signals: this is the case of frightened frogs waiting in the bottom of a pond. The delay till the end of alert seems to have a random value.

3.2.2.5.3. Self Referring Regulation of Drives Intensities

The working of the thirst drive device is clear except that we lack a reference value: there is not any sensor able to measure directly the need of water (its relation with osmotic pressure is masked by the storage of water). Then, how to determine the reference value? In most of cases, inborn value would be maladjusted. We know only one elementary phenomenon leading to a solution: the long term potentiation (Section 4.1). The yield of the synthesis of a transmitter is slowly adapted to the usual activity of a synapse. Then, the critical number of gulps would be a mean of our drinking habits during the last weeks.

Many other instincts use self referring control of the drive intensity, for instance sexual appetence, sleep duration, anxiety level.

Bad regulations are often observed. We have just described urge incontinency. Old people "forget" to drink. Many people are exceedingly anxious. Some drugs act on specific synaptic receptors to change the sensitivity of some neurons. After several weeks, the regulation is expected to regain its normal level. A more natural way is the building up of a learned inhibition. Thus, training leads to the inhibition of some effects of fear in actors or warriors.

3.2.2.6. Emotional Content of End Signals

3.2.2.6.1. Satisfaction and Pleasure

We have seen that the satisfactory ending of a drive (drinking when we are thirsty) triggers a periodic neuron. Its salvo inhibits the drive. In many cases, when the drive was sufficiently strong, the satisfactory ending (drinking when we are very thirsty, beginning to void after an urge sensation) triggers an impulse transmitted not only to inhibiting neuron, but also to a modulating pathway. In 4.2, we will see that this strong impulsive modulation is able: a) to allow the memorization of the circumstances; b) to lead to the repetition of a sequence of acts. It is often felt as a conscious pleasure. We will say that the end of the drive generates a **pleasure impulse** (Fig. **3.10**).

The pleasure impulse is transmitted to various cortical areas and also to the limbic system. It has been recognized that excitations of the septum and the hypothalamus are associated with pleasant sensations. In a dramatic experiment [24], rats could stimulate their septum by pushing on a pedal therefore inducing a pleasant (sexual?) sensation. They pushed till 7000 times in an hour and died from tiredness.

Note that electronic games, because bright and moving pictures on the screen attract the visual choice and because they induce impatience and pleasure impulses, although they do not have any functional usefulness, are built to trap the attention. We will see in part IV the leading role of the pleasure impulse: the limbic system sends then to the cortex a modulating signal necessary to allow learning.

3.2.2.6.2. Disappointment and Resentment

When impatience induces too long or too intensive a signal without reaching success, an end signal, which we call disappointment, stops the drive. Most of times, disappointment generates a strong modulating pulse, symmetric of the pleasure pulse and perhaps associated with the amygdalae. We call it resentment.

3.2.2.6.3. Weakly Satisfactory Acts

Any signal even without any relation with the causes of the drive can be linked with the drive system and acts to inhibit the drive and therefore is a satisfactory signal. Most of these arbitrary satisfactions are only weakly satisfactory, they stop only for a time the stronger drives. For instance, to yawn when you are drowsy, to cough when your throat is irritated or to shout out an oath when you are angry are weakly satisfactory acts.

3.2.2.7. Fragmented Instincts

Many great animal instincts (with the usual meaning of the word instinct) and many learned ones are sliced into a sequence of partial generalized instincts. Each one owns a specific drive. The satisfactory signal of a partial instinct triggers the impatience of the next one.

As an example, look at the feeding instinct of a cat. It is made of three generalized instincts: watching, catching and eating. The fragmented nature of such instincts appears when the sequence is not carried in the natural order. If a cat is well nourished by its master, its eating drive is inhibited by a satisfactory signal, but the watching and the catching impatience continues to slowly increase. Catching is suddenly chosen: during about ten minutes, the cat plays with the curtains or with a paper ball.

Even if carried out in the natural order, many fragmented instincts exhibit a ritual aspect. For instance, sexuality leads to the ritual sequence of approach, parade and so on. This ritual appearance is easily detected in many of our learned habits.

3.2.3. The Evolutionary Development of the Inborn Instincts

Main results: new inborn instincts are derived of old ones by a sequence of additions and of separations.

The diversity and sophistication of the inborn instincts are increasing all along the Darwinian Evolution. These progresses are probably due to a multi-steps mechanism. First, a new behavior is added to the original one, both of them triggered by the same order. For instance, a threatening grimace comes with the attack.

In the next two steps, the triggering of the new behavior splits off the old one and the signal accounting for the accomplishment of the new behavior makes links with the drive neuron and inhibits it: the accounting signal is become an (often weak) satisfactory one. Observe a dog facing up with a beggar. The intensity of the aggressiveness drive depends on the closeness of the beggar. The original behavior is attack, the grafted one is barking. In the Appendix, we present a plausible circuitry generating barking with a frequency increasing when the beggar gets closer and attack when he is sufficiently close to the dog.

The last step is the replacement of the signal initiating the drive by a new one. Karl Lorentz [25] showed that the greeting of the ash-colored goose is derived from threatening by substituting recognizing a friend for recognizing an enemy as initiating signal.

To sum up: new instincts are derived of old inborn ones by a sequence of additions and of separations. The same applies obviously to learned instincts. This observation will be very useful when we will study learning (in part IV).

APPENDIX: THE COMMAND OF BARKING

The circuitry (Fig. **3.11**) begins by a driving neuron firing continuously. Its frequency F increases like $1/d$.

In the inborn attack pathway, a slow synapse generates a command potential proportional to F. The attack neuron is a periodic one weakly coupled to the driving one. Thus, it emits a salvo if F is greater than some threshold (if d is weaker than some critical distance). The grafted barking pathway: it uses a generator of repetitive signals (see appendix I-J). The coupling with the driving neuron is strong; then the barking critical distance is greater than the attacking one. The driving signal modulates the device activity: the barking frequency increases when the beggar comes closer. Barking is inhibited during attack.

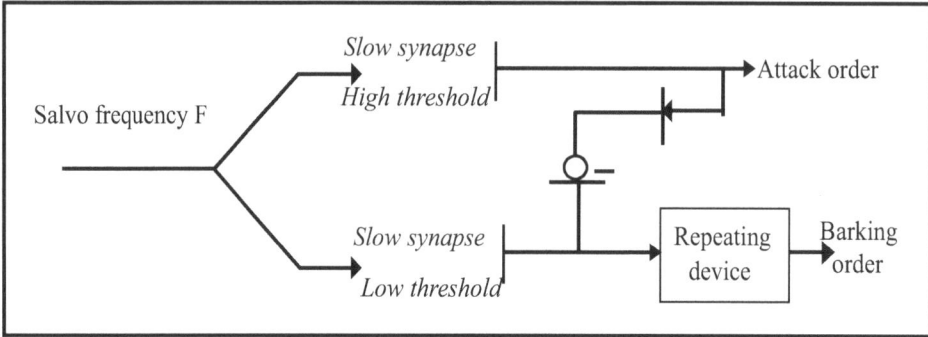

Fig. (3.11). How a dog chooses between bark and attack.

3.3. The Mechanisms and Control of Motion

The motions are controlled by a pile of hierarchized devices. The lower level one is built around a muscle. It is able to lift a weight and to keep it at a constant height. We will see that the unit is made of a muscle, an antagonist muscle (or several) and their embedded sensors. Each of these components had been described separately, but the working of the whole unit has been poorly studied. So, we will carefully describe it. The intermediate hierarchical device (the so called reptilian brain) is able to control walking and adapt the breathing rhythm to it. It is made of the spinal cord and the brain stem. The higher hierarchical device is supported by the cerebral cortex coupled with the thalamus and the limbic system. It chooses a goal (to go from my home to my office) and deducts from it a complex sequence of simple motion orders. Cerebellum and brain use trial and error learning. This will be the subject of chapter 4.2. Cerebellum is a large auxiliary device in charge of the stability of the body and of the accuracy of the motions.

3.3.1. The Muscle and the Lower Hierarchical Device

Main facts: Evaluation of the best muscular tonus Functions of the muscular embedded sensors. Need for phasic excitations.

Nature made up only one solution to transform chemical energy (the most usual fuel is ATP) into motion and mechanical work: all biological motions inside a cell (variation of the rigidity of the cilia of hair cells, active transportation of proteins along an axon) and all the global motions of cells (swimming of spermatozoa, contraction of smooth muscles, not studied here, contraction of striated muscles, our subject) seem to be due to the crawling of actin (or actin like) proteins on a myosin (or myosin-like) support.

The result is such an imperfect motor that its use requires length limitation and recourse to push pull devices.

3.3.1.1. Force Versus Nervous Order

3.3.1.1.1. The Muscle Components

A skeletal muscle is made of ligaments, tendons, contractile fibres, nervous cells and sensors. In a first approximation, tendons are described as rigid components. Nervous cells and sensors will be described in the next Section. The present Section is devoted to the description of the contractile apparatus.

A skeletal muscle is made of several motor units. A motor unit is a bundle of identical contractile fibres. (So, the force of a motor unit is proportional to its number of fibres). Each motor unit is excited by one and only one α motoneuron.

Motoneurons are slow neurons, with a threshold U_{THR} = 7.9 mV and frequencies ranging from 0 to 60 Hz. Tonic motoneurons are not transient. Phasic motoneurons are transient. After a delay of 200 ms, they would exhibit an asymptotic firing rate F_∞ with a range 0 to 5 Hz if not inhibited by some secondary circuitry. The synapses between motoneurons and muscular fibres are swift ones.

The logistics of the fibres are adapted to their function. Phasic fibres, which work only for short time, are fast fatigable: contraction wastes (mainly lactic acid due to the glucose consumption when regenerating ATP) are slowly eliminated. When these cells are excited too long a time, the wastes pile up and decrease the contractility. Tonic fibres, made to work in a continuous way, are fatigue resistant: they eliminate efficiently the wastes. One knows also moderately fatigable fibres.

3.3.1.1.2. The Mechanism of Contractility

Fibres contain actin and myosin filaments linked by troponin C molecules, which are activated by the firing of the motoneuron.

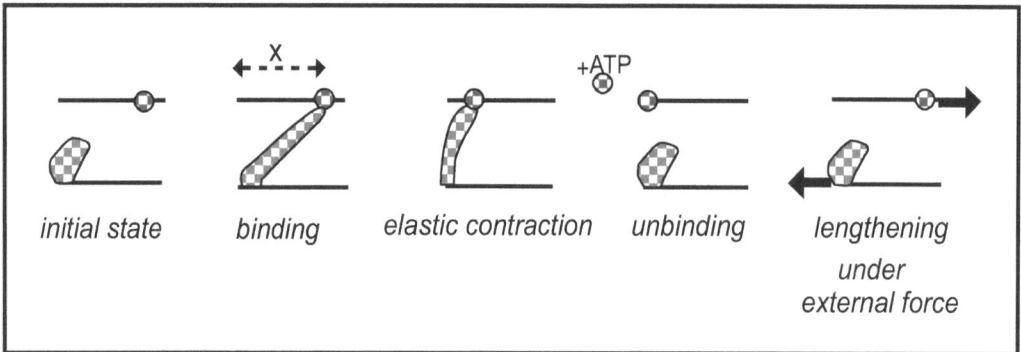

Fig. (3.12). Contractility results from the actin-myosin binding and unbinding.

When they are activated, actin crawls upward along the myosin while it is pulled downward by the external force (the load). Huxley's sliding filaments model [26] describes quantitatively the phenomena (Fig. **3.12**) and Appendix A).

The Energetic Point of View: Each elementary cycle consumes one ATP molecule. If the muscle is loaded, each little contraction is followed by a little shift backward and the total muscle length does not change: the mechanical work performed by the muscle is zero while the consumption of ATP is proportional to the time. So, there is a constant power consumption (power = energy/s) without mechanical energy production.

3.3.1.1.3. Quantitative Study of the Muscular Force

Force Versus Muscle Length: The force is maximum if the actin and the myosin filaments are facing each other. This is not the case if the muscle is distended (Fig. **3.13**, left) or too contracted (Fig. **3.13**, right).

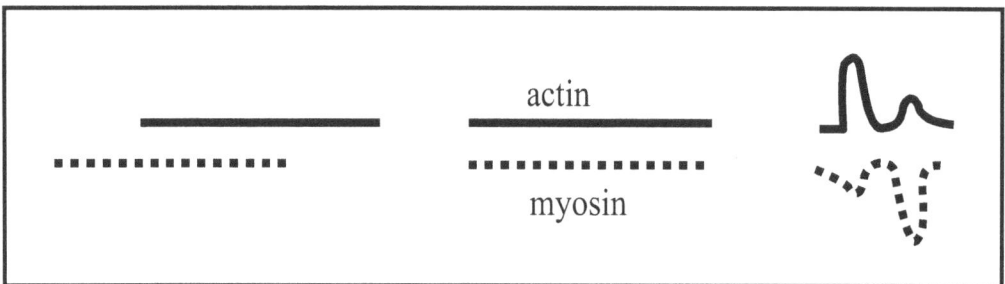

Fig. (3.13). Excessive or weak muscle lengthening decrease the contractile force.

This effect is expressed by a gaussian length factor

$$\alpha = \exp-\left(\frac{L-L_0}{0.4L_0}\right)^2 = \exp-\left[6.25(1-y)^2\right]$$

where L_0 is the reference length of the muscle (the length allowing the greatest force) and y the reduced length L/L_0.

Force Versus the Lengthening Velocity: We have seen above that the Huxley's model gives a velocity factor (Fig. **3.14** and appendix A).

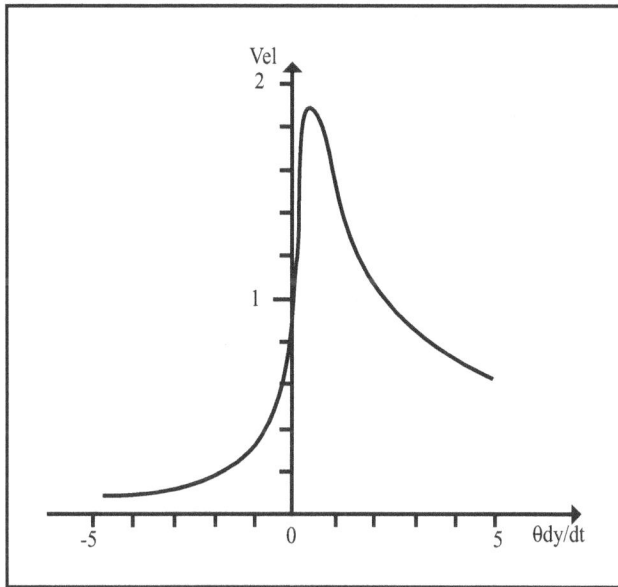

Fig. (3.14). The velocity factor (y is the reduced length of the muscle L/L_0 and θ is a characteristic time scale).

Remark the strong resistance to moderate lengthening velocities (the factor Vel of the contracting force reaching the value 1.9), then the submission to more powerful lengthening. Note that the well known empirical Hill's law [27], which fits well the experiment during moderate contraction, does not apply to lengthening (it would give an infinite force).

The Progressive Recruitment of Motor Units: Motor units can be numbered according their size (their number of fibres), starting from the weakest. Experiment shows a) that the number n_K of fibres of the k unit is an exponential function of k (Appendix C); b) that, when the unit k is firing with a moderate firing rate, all the units k' with k'< k fire at the maximum firing rate while all the units with k'> k are mute. (Strong excitation of all neurons k'< k by the neuron k leads to this property). Thus, the number of working fibres increases in a progressive and well arranged way.

The Force of the Intermediate Motor Unit k: This motoneuron k has a firing rate F_K such that $0 < F_K < F_{KMAX}$. Each spike of the motoneuron induces the entrance of Ca ions into the fibre, which activate the Troponin C. From the study of this cycle [28] we deduce (Appendix B), that the force is roughly proportional to the firing rate F_K.

Summing Up: A striated muscle is characterized by its reference length $\mathbf{L_0}$ and by its maximum force $\boldsymbol{F_{MAX}}$. The usual Hill's law contains a critical contraction velocity b, which was measured for most of muscles. From b, we deduce the characteristic time $\boldsymbol{\theta} = L_0/b$, which defines the time scale of the muscular contraction. The number of motor units and considerations about noise (see 1.3) define the exponential factor ruling the motor units' maximum forces. Then, the total muscular force can be written:

$$F = F_{MAX}. \, \alpha(y).\text{Vel}(\theta dy/dt).\Phi(\text{recruitment})$$

3.3.1.2. Searching for Natural Stability: The Muscular Tonus

A muscular device supporting a constant load and excited by a constant nervous order (thus without phasic excitations, which cannot be constant) is naturally stable if it keeps a constant length and return to its equilibrium length after a small length perturbation. Note that a stable device obeys to slowly changing orders. (Natural stability differs from forced stability, which is obtained when a loop changes the nervous orders to compensate a perturbation).

3.3.1.2.1. Skeletal Joints have to Limit the Range of Possible Muscle Length

The joints mechanically limit the possible lengths of any muscle. The ligaments are elastic security devices, which stop the motion when it reaches a limit of the range.

We assume that this limit is a functional optimization. Look at a muscle (relative length y, excitation Φ) supporting a mass M. If it is too lengthened (y > 1), a small lengthening perturbation causes a decrease of the contractile force (because $d\alpha/dy < 0$):

the system is naturally unstable; the values y > 1 have to be forbidden. On the other hand, if the muscle is excessively contracted, the length factor α is small and the force very weak: α = 0.5 if y = 0.65. So, we assume that the joints limit the possible lengths y to the range [0.65 < y < 0.95].

3.3.1.2.2. An Isolated Muscle Cannot be Controlled

This property is obvious when the load M is very weak: Then, if the recruitment factor Φ is zero, the muscular length y is undetermined, the muscle would be flaccid. If Φ is not zero, the contractile force is unbalanced; y would tend toward its minimum authorized value. Thus, antagonist muscles are needed to stabilize the agonist ones.

3.3.1.2.3. Looking at an Academic Two Muscles Device

To avoid the difficulties of a realistic description which would look at the path of antagonist muscles around a joint (it is not our subject), we look at the oversimplified schematic device of the Fig. (**3.15**).

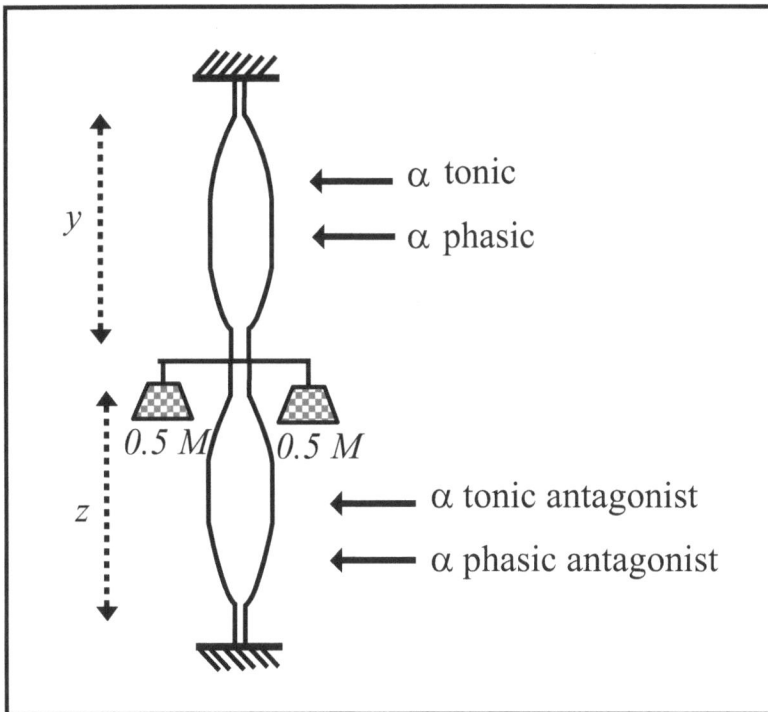

Fig. (3.15). A simplistic scheme with a load and two antagonist muscles.

Although strongly unrealistic, this model furnishes qualitative information about the role of tonus, spindles and of phasic excitations.

We assume identical characteristics for the two muscles (and we choose as an example the characteristics of the biceps femoris (same individual maximum force F_{MAX}, $L_0 = 40$ cm and $\theta = 0.1$ s).

The two reduced lengths y and z are linked in such a way that both of them remain in the governable range $0.65 < y, z < 0.95$. Hence: **y + z = 1.60**

The motion equation is obtained from the Newton's law:

$ML_0 \, d^2y/dt^2 = Mg +$ antagonist force $-$ agonist force (with $g = 981$ cm/s^2)

(Without muscles, the load freely falls). This equation allows studying the equilibrium conditions and the response to little perturbations (Appendix D).

Defining the Tonus at Equilibrium: The intuitive notion of tonus has to be specified. We define the tonus T as the mean force of the two muscles compared to their maximum force:

T= ½(Agonist force + antagonist force)/F$_{MAX}$

(range: 0, 1). Note that T is proportional to the power expended by the device (the number of ATP molecules consumed each second by the two muscles system). This energetic definition of tonus is still valid during motion.

Equilibrium Conditions: A tonic muscle with maximum force F_{MAX} cannot lift a weight greater than $g.M_{MAX} = \alpha(y).F_{MAX}$ (remind that α is about 0.5 for y_{MIN}).

Return to the Equilibrium Position: After a little perturbation δy, y returns to its equilibrium value like $\exp(-ft).\cos(\omega t)$ or like $\exp(-f_1 t)+\exp(-f_2 t)$ with $f_1 < f_2$. The return is slower for a lengthening ($\delta y > 0$) than for a contraction. Engineers use a rule of thumb: the **return time** RT needed to return at the equilibrium position after a small lengthening is of the order of magnitude of $RT = 4/f$ or $4/f_1$. This time is easily evaluated (Appendix E).

3.3.1.2.4. The Most Desirable Tonus

Searching for a Criterion: In this paragraph, we do not look at the nervous device able to build such a tonus, but only at the goal to be reached. For any load and any equilibrium length, we find that the return time decreases when the tonus T increases. But a) very short RT lead to unpleasant overshoots and b) great tonus are energetically expensive. (The ATP consumption is proportional to T. Without antagonist muscle, the equilibrium condition leads to $T_{MIN} = 0.5 \ Mg/F_{MAX}$. Thus, **the excess power** (EP) due to the antagonist muscle is $EP = T - 0.5.Mg/F_{MAX}$). Accordingly, we search for a moderate RT. The time scale of the muscular motions is the characteristic time θ; we assume that the best excitations (and the best tonus) must lead to $1/f = \theta$ (RT = 4θ).

An Unexpected Result: For each muscle length y, each load M (or $\mu = Mg/F_{MAX}$) and each tonus T, we easily compute all the equilibrium parameters: excitations Φ and Ψ (that is the number of recruited muscular fibres) of the agonist and the antagonist, numbers k and k_A of recruited motor units (the excitations are exponential of k), excess power EP, length factor α and (since μ must be smaller than α to satisfy the equilibrium conditions) the ratio μ/α (range 0, 1) and characteristic parameters ω and f or f_1 and f_2 of the potential oscillations. Then, we try any kind of plot, searching for simple properties (with in this example $\theta = 0.1$ s).

We obtain values 1/f conveniently close to θ by assuming a linear excitation law of the agonist muscle if $\mu > 0.5\alpha$: $\Phi = 2\mu/\alpha$ for $\mu/\alpha <0.5$ and $\Phi=1$ for $0.5 < \mu/\alpha < 1$.

The unexpected result (Fig. **3.16**, left) is that we obtain a very simple plot of 1/f *versus* the ratio μ/α and that this value 1/f does not depend on y.

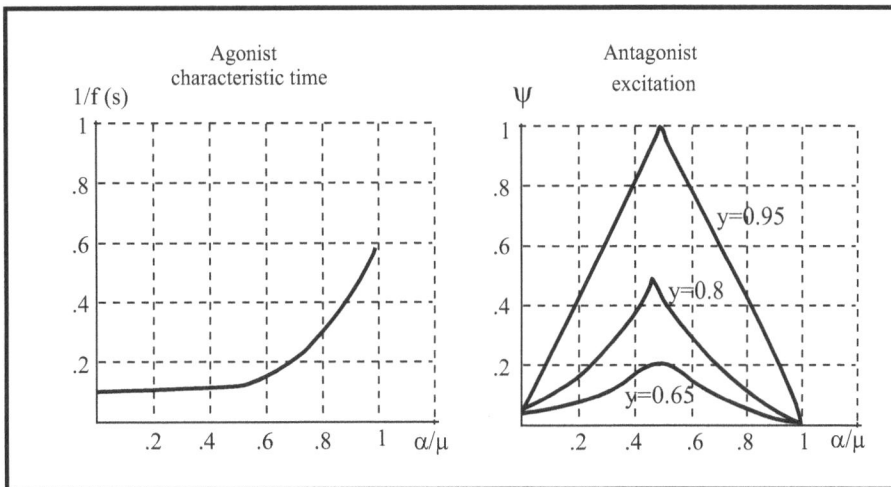

Fig. (3.16). Agonist and antagonist best tonus as a function of the load.

If $0 < \mu < 0.5\alpha$, 1/f is almost a constant close to 0.08 s (while $\theta = 0.1$ s).

If $0.5 < \mu/\alpha < 1$, there is no equilibrium with $1/f \approx 0.1$ s. That means that heavy loads induce slow muscular reactions. We computed then the smallest 1/f value. It is obtained when the excitation of the agonist muscle is maximum ($\Phi = 1$).

If $\mu > \alpha$, there is no possible equilibrium. The excitations of the agonist and the antagonist muscles do not play symmetric parts. While Φ does not depend directly of the length y, the curves Ψ *versus* μ/α depend strongly on the y (Fig. 3.16, right). If $\mu/\alpha < 0.5$, an increase of the load with constant y causes an increase of both Φ and Ψ. If $\mu/\alpha > 0.5$, Φ keeps its maximum value $\Phi = 1$ and an increase of the load causes a decrease of the antagonist excitation Ψ.

These properties of the muscle will be a key for understanding the nervous control of the muscles.

3.3.1.3. The Implied Sensors

The skeletal mechanics are very intricate. For instance, hand prosthesis usable for teeth brushing must have at least 6 freedom degrees (6 rotation axes); its control needs at least 6 sensors. More generally, the control of posture and motion in man requires a great number of sensors of various kinds. We have to describe any of them. Another problem is to gather their huge and ill-assorted data to build proprioception.

Most of them have the cell body and the axon of a standard neuron while the input channels are opened by mechanical deformation, for instance hairs are opened by bending (Fig. **3.17**). The rigidity of the hair cell controls its functional sensitivity. In the most optimized case (the threshold of hearing), a 0.3 nm lateral motion of the hair cell products the opening of a 0.3 nm wide K channel.

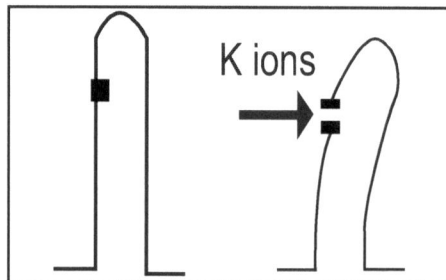

Fig. (3.17). Mechanism of hair cells sensors.

Hair cells are the transducers for hearing, head inclination and rotation.

3.3.1.3.1. Ruffini's Joint Sensors

We have seen (1.3) that these non transient sensors arer adapted to the measurement of the knee position.

3.3.1.3.2. The Head Kinematics

A kind of gelatinous pendulum supporting a high density ($\rho = 2.5$) rigid plate is bended by the head inclination. It measures the angles between the vertical and the head axis. (Replacing the high density crystals of this system by ferromagnetic crystals, we obtain the sensor of the geomagnetic sense of the pigeon).

3.3.1.3.3. Head Rotation

Three "semi-circular channels" (one for each axis) bound to the head are full of fluid and partially obstructed by a gelatinous barrier (the cupula) in which hair cells are embedded. When the head is submitted to a rotation, the inertia of the fluid induces a bending of the hair cells (Fig. **3.18**). Transient or non transient cell bodies measure the head acceleration or the head velocity.

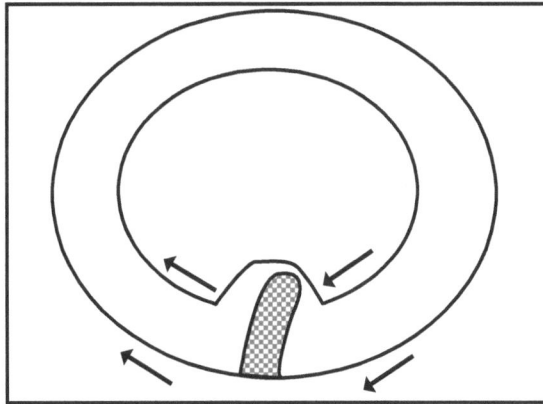

Fig. (3.18). Internal ear accelerometer.

3.3.1.3.4. Skin Mechano Sensors

They measure the pressure exerted on the skin to report first contact (transient cell body, used to end a motion) or permanent contact (non transient cell body, used to control holding).

3.3.1.3.5. Miscellaneous Sensors

We have to deal with pain sensors and temperature sensors. Do not forget vision.

3.3.1.3.6. Embedded Muscular Sensors

Three types of sensors, mechanically bound to the muscular fibres or to the tendons, are embedded in each muscle. Two properties make them remarkable:

a)   Their number is very great, allowing a progressive recruitment of their outputs (1 tendon organ for 20 contractile fibres, more than two β or γ motoneurons for one α motoneuron).

b)   They are not simple sensors, but more precisely comparing devices owning two inputs: they are excited (or inhibited) for one part by specialized motoneurons (β, γ_{TON}, γ_{PHA}) and for the other part by ionic channels the opening of which is due to the muscle contraction. Their body is very similar to a neuron cell body. Their spikes are transmitted by long axonal fibres (I_A, I_B, II), which reach the medulla. (Fig. **3.19**). Thus, their control potential (and their output firing rate) is proportional to the measured value of some muscle property (length, force…) minus the wanted value of this property.

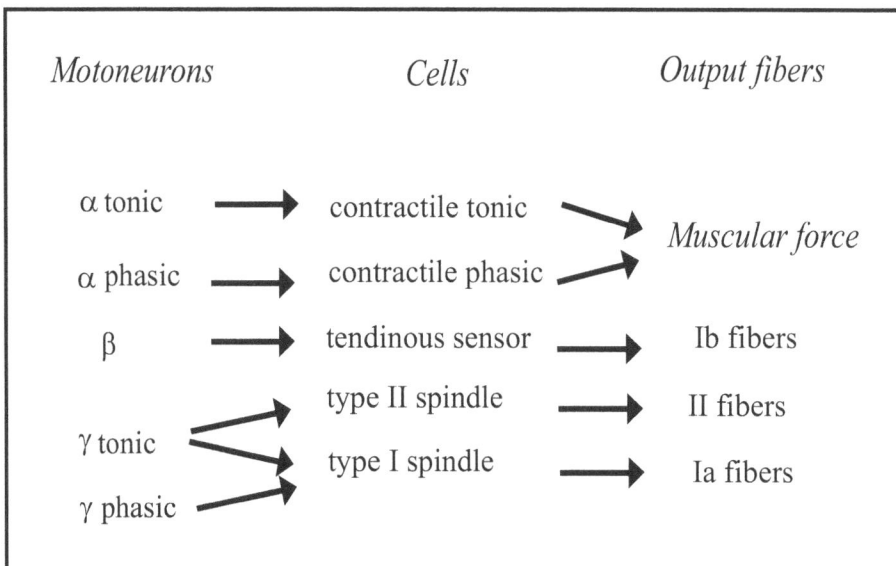

Fig. (3.19). Connections between the motoneurons and the output fibres.

3.3.1.3.7. Modeling these Sensors

The Method: The experimental observation of the cells properties [29, 30], and the knowledge of the general properties of neurons lead to build up simple models of the sensors and to verify that these models agree fairly well with some purely empirical models, which have already been published [31-33].

Type II Spindles: At each time, their length y is the same as the length of the contractile fibres. In addition, they are inhibited by γ tonic motoneurons, (which are standard slow neurons). The spindle behaves as a linear neuron. Thus, the firing rate of the II fibres (the output) is proportional to $y - y_{WANTED}$ (Appendix E).

Golgi Tendon Sensors: They are embedded in the tendon, which is an elastic string (its strain is proportional to the total muscular force). Thus, the opening of their ionic channels (which has an inhibiting effect) depends on the muscular force F. Exciting inputs are transmitted by β motoneurons. Golgi sensors act behave as neurons, the axons of which are fibres Ib. Thus, the firing rate of the Ib fibres (the output) is proportional to the difference between a wanted force and the measured one (Appendix E).

Type I Spindles: The two ends of such a spindle are bound to the muscular fibre. Thus, its length is proportional to the muscular length y. But the spindle is not homogeneous (Fig. **3.20**). It is made of an elastic part (length x) and an elastic and contractile fibre (length z), the contraction of which is governed by γ phasic motoneurons. The elastic part is similar to an ordinary (slowly transient) neuron cell body. Its excitation is due to the opening of its ionic channels; it depends on the length x (and not on the total length y: with constant y, x increases if the γ phasic motor is firing). The firing of the γ tonic motoneurons inhibits it.

Experiment shows that if the muscle is suddenly lengthened, all the Ia fibres are recruited and act directly on the phasic α motoneuron, causing a brisk contraction of the muscle, the well known stretch reflex. The spindle model (Appendix E) describes conveniently the fibres firing. In a rough approximation, fibres Ia generate a short burst of high firing rate spikes if $y - y_{WANTED}$ is greater than some threshold and the γ phasic firing control the value of this threshold.

3.3.1.4. The Control of Muscular Tonus

Tonic orders do not act directly on α motoneurons, but through the local control device. This device is convenient to control the response to slowly varying (in more than 0.2 s) length orders or loading. But the response to a brisk change needs phasic excitations.

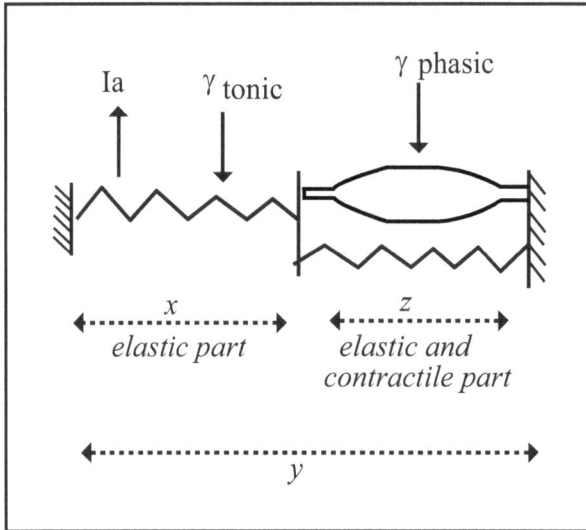

Fig. (3.20). Structure of Type I spindles.

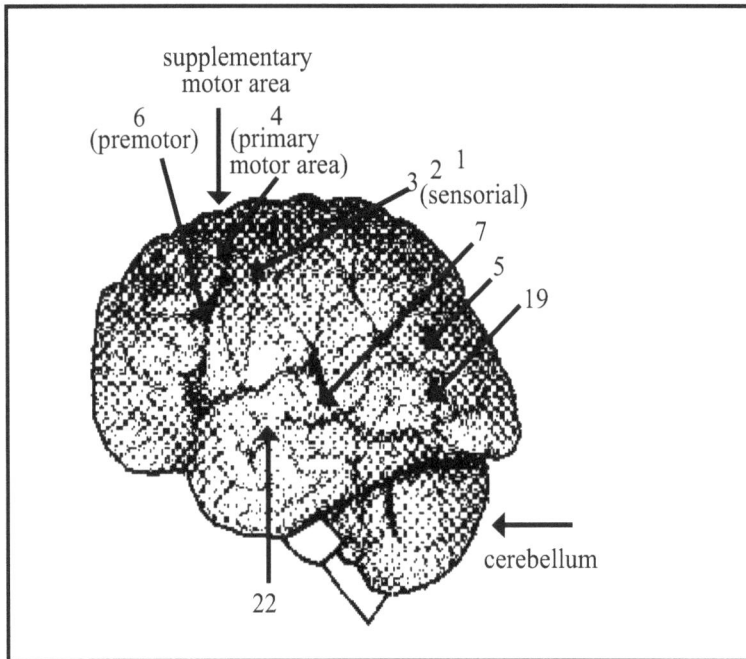

Fig. (3.21). The main cortical motor areas.

3.3.1.4.1. A General Scheme

We look at the tonic equilibrium of a loaded system made of two antagonist muscles, their embedded sensors and an associated spinal node (Fig. **3.22**). We assume that the signals of the other sensors are not transmitted to this device with the exception of pain signals which act on the phasic fibres but not on the tonic ones.

There are 10 types of motoneurons going from the spinal node to the muscles (tonic α, phasic α, β, tonic γ, phasic γ for each muscle) and there are 6 types of sensor fibres going from the muscle to the spinal node (Ia, Ib, II for each muscle). In the case of a purely tonic behavior, the part of phasic γ motoneurons and Ia fibres can be neglected. The node receives orders from the higher hierarchical levels and sends them data. Our goal is to specify what orders and what data.

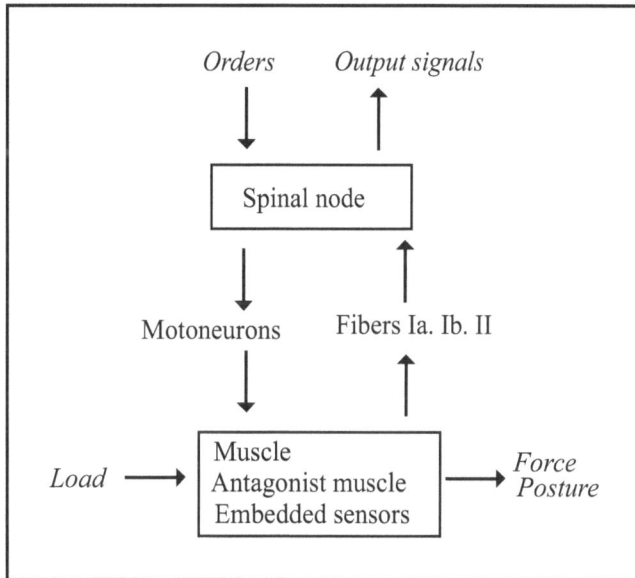

Fig. (3.22). The nervous circuitry.

3.3.1.4.2. The Orders

Slow fibres of the pyramidal tract (Appendix F) excite the β and tonic γ motoneurons. They probably do not excite directly any α motoneuron. We assume they transmit a muscle length order y_{WANTED} and an overloading prediction $\mu_{PREDICT}$ (which will be competitive with the actual load μ_{ACTUAL}). Medulla injuries induce extreme (very small or very high) values of the orders y_{WANTED} and $\mu_{PREDICT}$.

3.3.1.4.3. The Outputs

Two meaningful messages are computed (Appendix F): actual load $\mu_{EFFECTIVE}$ and difference $y - y_{WANTED}$ between the actual length and the wanted one.

3.3.1.4.4. A Plausible Circuitry

We start from the following remarks: a) Simple experimental tests (trying to lift an unexpectedly light or heavy object) show that, in many cases, we adjust the tonus before starting the motion. More sophisticated experiments (electromyography of both a muscle and its antagonist, study of the action of the premotor cortical area…) confirm this result: the adaptation to load and the length are governed by separated loops. b) The load control is measured by the Golgi tendon sensors, the muscular length by the type II spindles. c) The device must impose an equilibrium tonus close to the best one. Then, the agonist excitation has to be controlled by the load (and the antagonist by the length). d) The progressive recruitment of motoneurons α is more closely related to the number k of firing neurons than to the total force (an exponential function of k). On the other hand, the great number of sensors allows a progressive recruitment of their outputs, leading to a measurement of $Log(\mu)$, a more sensitive signal than the brute load for weak values of μ. And the best relation between $Log(excitation)$, $Log(\mu)$ and $Log(\alpha)$ is a linear one, easily built with neurons.

These remarks lead to the tonic part of a simple model. A phasic part has to be added. (Appendix F). Satisfactory results are described hereafter..

3.3.1.5. Fast Motions

3.3.1.5.1. Fast Response of the Two Muscle Device

Phasic control is assumed by the type I spindles and by Renshaw's neurons (Fig. **3.23**). Renshaw's neurons are transient swift neuron excited by the phasic α motoneurons; In return, they retro-inhibit the α phasic motoneurons, letting them to fire during less than 30 ms (a time fitting well the duration of the motion acceleration phase). Add to this that the Renshaw's neuron can be inhibited by pain transmitting neurons (fibres III) and excited or inhibited by higher level orders. Experiment shows that the outputs of the type I spindle have a monosynaptic link with α phasic motoneurons. Thus, a fast pain signal (fibres III) induces an immediate contraction of the muscle. The same device induces a contraction when the muscle is suddenly lengthened: it is the well known stretch reflex.

Note that the need of cross linking between tonic antagonist and antagonist phasic circuitries: antagonist Renshaw's neurons have to be excited during a wanted contraction to avoid an automatic braking due to the stretch reflex.

The local control adds phasic agonist and antagonist orders to the tonic ones when acceleration and deceleration of a purely tonic motion are needed. This assumption is confirmed by very old experiments [40]: the electromyographic activity of both an agonist and an antagonist muscles were recorded during various motions. Slow motions (without overshoot) were governed by a continuous agonist excitation, medium velocities (with only one overshoot) by a long agonist excitation followed by a short antagonist excitation, great velocities (with an oscillating behavior) by a succession of short agonist and antagonist impulses.

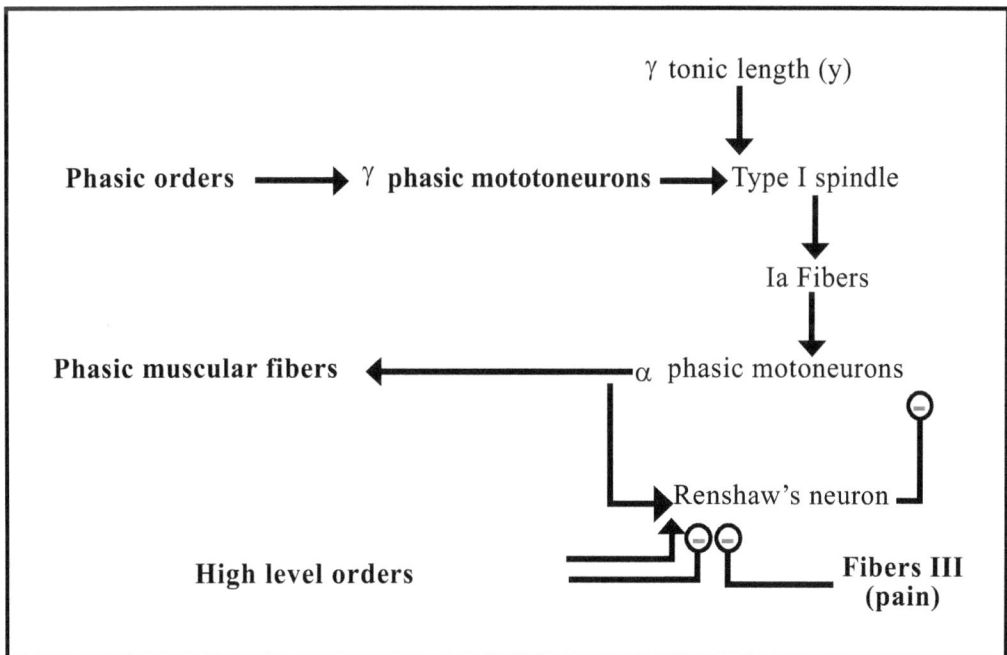

Fig. (3.23). The phasic circuitry.

3.3.1.5.2. Computational Results

The scheme of Appendix F leads to satisfactory results. Thus, a local model with tonic and phasic control gives satisfactory results. Without phasic control, y would reach unstable zones and so could not be controlled. A predicting-correcting control (in charge of the cerebellum) is required to avoid overshoots.

We show hereafter the muscular length *versus* time for y_{Wanted} changing briskly from 0.9 to 0.7 with a weak load $\mu = 0.2$ (Fig. **3.24**, left), and for a brisk loading (right): with $y_{WANTED}=0.7$, $\mu = 0.1$ for $t<0$ and $\mu = 0.5$ for $t > 0$. The computed firing periods of the phasic neurons are indicated.

Muscular Overpower: In some situations (striking a tennis ball, lifting weights), the cortex excite strongly the γ phasic motoneurons of the agonist and inhibits its Renshaw's neuron: then, the phasic contractile fibres are used as if they were tonic.

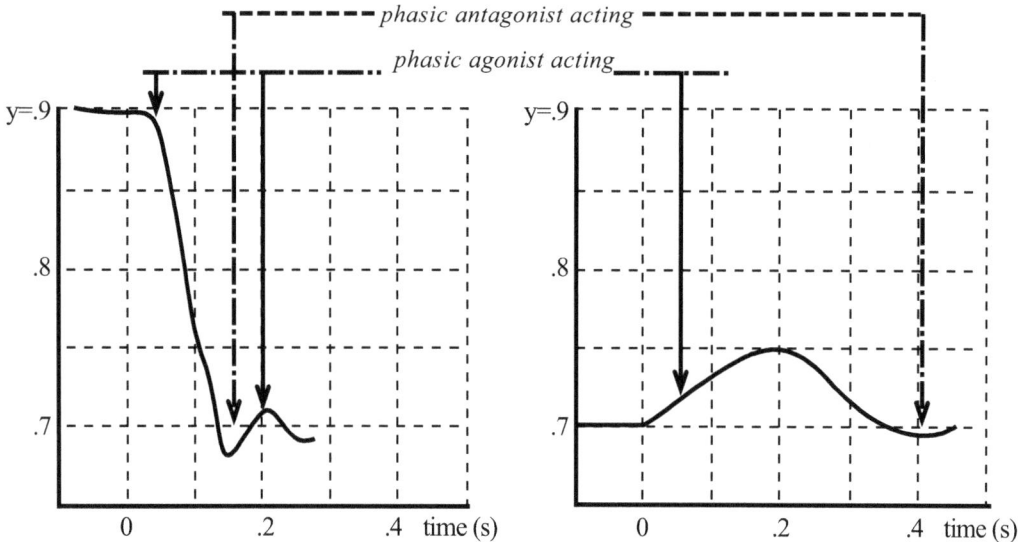

Fig. (3.24). Computed response to the brisk contraction order (weak or heavy load).

3.3.2. The Intermediate Device

Main facts: Spinal automatisms. Control by the reticular formation of automatisms such that breathing.

3.3.2.1. The Reptilian Brain

The intermediate device is made of the spinal cord and the brain stem which makes the junction between the spinal column and the brain. It contains centers controlling the autonomous nervous system. It governs for instance the vomiting reflex. The reticular formation is a set of inter-meshed neural networks inside the brain stem. In the brain stem, we found also the pons, which is a relay station carrying signals from cerebral cortex, eyes, ears and touch receptors to the cerebellum.

Some people call "reptilian brain" the intermediate device. All newborn mammals seem ruled only by the reptilian brain. Look at the example of a baby calf. It stands up, moves forward if slightly pushed, it finds the udder if brought close to its mother and immediately sucks it. It does not see anything; it is unable to choose behavior. It lacks brain to be a fully operational animal.

3.3.2.2. Spinal Automatisms

The spinal cord is not only a double transmitting bundle of ascending and descending fibres (Appendix F), but also an almost autonomous control system.

Sherrington has shown [38] that decerebrated cats and dogs were still able of some complex movements, which implies obviously a global coordination of the limbs due to the spinal neuronal circuitry. If one causes an itch on the breast of a decerebrated dog, it scratches itself with its more suitable leg. If one immobilizes this leg, the dog scratches itself with another leg. Other example: if one puts down a decerebrated cat on a moving walkway, the animal walk or trots or runs in accordance with the walkway velocity. The spinal automatism generates both tonic and phasic orders. Experiments show that each limb owns in the medulla a pace maker (called central pattern generator) and that coupling between the four pace makers determines the limbs synchronization and the choice between walking and running. On the other part, a contact on the front of its paw makes the cat lift higher its leg as in the face of an obstacle. (The automatic walking of the newborn baby seems a similar phenomenon). In man, several spinal automatisms have been described, for instance the response to a local burning of one foot: the fast of the stimulated leg is accompanied by an increase of the tonus of the extensor muscle of the other leg. The structure of the operating neuronal circuitry is easily guessed.

3.3.2.3. The Reticular Formation of the Brain Stem

Since the pioneer work of Magoun and Moruzzi [39], we know that the reticular formation is in charge of the control of nervous orders which (as the breathing rhythm) can be modulated by the cortex or the circumstances, but are widely autonomous. A typical example is the sequence of breathing in and breathing out orders. Analyzing the experimental results of electrophysiology, we have described (Chapter 1.3, Appendix I and Figs. (**1.28**, **1.29**) the neuronal circuitry at work. This circuitry is characterized by the opposition of two behaviors (breathing in and breathing out). The reticular formation contains similar centers controlling the choice between other sorts of opposite behaviors, for instance the choice between deep sleep or sleep with rapid eyes motions, between transmitting or filtering pain signals. And it controls the muscular tonus.

3.3.3. Control of Motion at the Cortical Level

Main facts: Representation of the body. Navigation. The part of imaginating devices.

3.3.3.1. Cortical Analysis of the Incoming Data

3.3.3.1.1. The Body Mapping on Cortical Areas

The raw data coming from the various kinds of sensors are expressed in several coordinate systems: tactile signals are related with body surface coordinates (the top of my index), joint signals with local angles (bending of the knee) and visual signals with the retina coordinates, which can be reported to the location of the body relative to the room Cartesian coordinates (height above the ground, horizontal distances). When we turn or tilt the head, the location of our body does not change but the retina coordinates are transformed. To use this ill-assorted set of information, brain has to put them in unified systems. This is probably the part of the first cortical areas.

After relaying in the postero-ventral area of the thalamus, the signals of the skin sensors reach the Broadman's cortical areas 1 and 3b. Signals coming from the embedded sensors and the Ruffini (joint sensors) reach the postero-ventral area of the thalamus, then the Broadman's cortical area 3a. (Fig. **3.21**). The vestibular signals (tilt and acceleration of the head) reach the area 3a while the area 3b receives the face signals. Signals from areas 3a and 3b are mixed in area 2.

3.3.3.1.2. Using the Sensorial Information

Translated and gathered sensor signals are processed to obtain 3 different usable signals:

Representation of the Body: The processing of these signals is analogous to the processing of visual signals (Section 3.1). In a first step, all the signals concerning the state of the body are mapped into body surface coordinates without information loss. The geometrical characteristics and the mechanical ones (the forces) are analyzed separately. From the first, the use of inborn "passwords" allows to build an outline of the posture. Then, learned passwords allow recognizing some typical situations. For instance, the combination (opened joints + weak muscular tonus + light pressure on all the surface of my back) is the signature needed to induce the firing of a specialized neuron. Its activity labels for (has the meaning): "I in my bed". In a second step (Section 4.2), a visual representation of the body posture is build up. The parietal associative cortex and the pulvinar (a part of the thalamus) seem to be in charge of the perception of the body posture and of its stimuli.

Representation of the Body within its Surrounding (Navigating with a Map): Visual signals furnish a description of the room surrounding me. The retinal coordinates of each object change for any of my motions. Thus, a first storable representation is "I in a moving surrounding". It seems much easier to define a strategy in coordinates linked to the room. After such a coordinates change, each object is fixed and the actual or imaginative "I moving" is a volume located in this reference space. The room arrangement is often memorized. Then, I can plan my displacements from the bed to the door and, during the night, I am able to move without light. In most of cases, we have at our disposal a memorized room map for each of our usual places. Recognizing the place induces the choice of the appropriate map. But if I awake in a new room, I keep remembering the arrangement of the old one, I am unable to plan my displacements and, as reported by Marcel Proust, I feel disturbed.

This short-range representation is often supplemented by one or several long-range representations: "me personally" (the building of this signal will be discussed in chapter 4.2) is taken as a movable point on a two dimensional topological map made of a set of beacons. Such a map allows going from home to some far place following a route and coming back by another route. Men and dogs are able to do so. Children do not own such a map and they are easily lost.

The parietal and the frontal cortex [34], the latero-posterior thalamus and the hippocampus seem to be in charge of mapping and navigation (rats [35] and men [36] have "place cells" in their hippocampus, which fire when their owner in some particular places).

Recognition of the Shape and the Texture of Objects Held in the Hand: The associative cortex seems to be in charge of it.

3.3.3.2. The Control of Motions

3.3.3.2.1. The Part of the Cortex

Look now at a squirrel (a very primitive mammal). A noise or a gesture of the observer alert it, it runs towards a tree, jumps and climbs the face of the trunk opposite to the observer. This is a typical instinctive sequence. The alert signal is a drive strong enough to be chosen. It triggers the flight behavior. With its little brain, our squirrel has to see, recognize and choose a tree, then to run toward it. Running is a spinal automatism, but the high level controls have to modulate it to impose the direction of the run. Thus, the control devices appear as a three components system: the first one uses sensorial signals (here, alert signals) to define an immediate goal (run to a tree); the second uses sensorial inputs to represent the animal within its surrounding; the third makes a synthesis of these data to trigger, modulate and orientate spinal mechanisms.

Squirrel owns only a small number of possible goals, all of them simple and most of them inborn. Human possible goals seem countless. Man is able to achieve new purposeful acts, to learn and make them automatic. His motor control is able to use learning to deeply modulate or to bypass the inborn spinal mechanisms. We will describe in chapter 4.2 the associative devices which command the cortical motor areas (completed by the cerebellum which computes fast phasic corrections to slow motions). Note that experiments have shown [37] that the number of firing neurons in the primary motor area of the cortex and the number of firing α motoneurons are roughly the same (so allowing a direct command of muscles).

3.3.3.2.2. Delicate Motions

Most of times, the brain creates new motions by distorting older ones. It can: a) trigger and modulate spinal automatisms. For instance, any change of direction of a squirrel is obtained by modulating the automatic run; b) modulate older high level automatisms: in man, pedaling on a cycle is obtained by modulating the walking automatism. Thus, each of his muscles work for several masters (the different behaviors). The hand motions, for instance writing, are often completely new.

The signals transmitted to the descending pathway are triggering orders, tonus orders, tonic muscles length orders and phasic excitation orders. The synthesis of these orders result from an intricate processing. In man, the upright (and still more upright on one foot) postures are naturally unstable. Stabilizing feedbacks are specially developed. One of them controls the coordination of the tonic muscles to maintain the gravity center above the sustentation base (for instance by moving arms when walking). Another builds up phasic orders to compensate any beginning loss of balance.

3.3.3.2.3. The Part of Imagination

The neuronal mechanism of imagination will be discussed in chapter 4.2. Coordination of several muscles is carried out by a sequence of imaginative trials defining real tonus orders and rough virtual motion orders (look at a cat working out its jump). When the fit with the immediate goal is obtained, the rough motion order is transformed into detailed muscle length orders and carried out.

3.3.3.2.4. Cortical Motor Areas

This description, based on the observation of behaviors, is comforted by electro physiology and by the study of the effects of localized lesions: different cortical areas are in charge of the imaginative and of the effective motor command (Fig. **3.21**).

Search for a Strategy: Orbito frontal associative cortex, medio-dorsal thalamus.

Imaginative Motor Trials: Premotor cortex (Broadman's area 6).

Coordination of Polymuscular Motions: Supplementary cortical motor area.

Execution of Motions: Primary cortical motor area (Broadman's area 4).

3.3.3.2.5. The Part of the Thalamus

The cortex uses a great number of swift transient neurons regulated to be summitted at rest at a control potential just above the threshold. The basal ganglia of the thalamus ganglia (corpus striatum, palladum, subthalamic nuclei, nucleus niger) are in charge of these regulations. Some intricate feedbacks control their activity.

The corpus striatum receives signals from the associative cortex and from the nucleus niger. The nucleus niger receives signals from the premotor cortex and from the corpus striatum. If the nucleus niger is impaired (Parkinson disease), the thalamus is under excited. The cortical neurons at rest are maintained at a control potential definitely lower than the threshold. So the specific excitations have to be greater than usual to trigger the firing of a neuron. Thus, body motions are rarer, slower and with a less than usual efficient regulation (shaking appears). When the corpus striatum is impaired (Huntington disease), the thalamus is overexcited; the control potential at rest of the cortical neurons is too high; some competitive inhibitions can be ineffective and unwanted motions appear.

3.3.4. The Cerebellum

Main facts: Artificial stabilization of erected station and accurate motions.

3.3.4.1. Usefulness of the Cerebellum

After a synthetic review of experimental facts, the cerebellum appears as the device generating phasic excitations of skeletal muscles.

3.3.4.1.1. The Need for Phasic Excitations

Studies of cerebellum dysfunctions show that the cerebellum coordinates body movements and contributes to its equilibrium. It is a center for learning motor skills, but its failure does not impair slow motions governed by tonic excitations. Look now at some precise examples. A frequently used test (touch with a finger a target located at one meter from the patient) fails in the case of cerebellum dysfunction. It is described by a device similar to the simplistic one studied in 3.3.1 (Fig. **3.24**) without load (M = 0). We have seen that the antagonist muscle needs a short phasic excitation at the end the motion. With only tonic orders, the finger would not reach the target or would crash into it. Phasic orders allow reaching the target with a zero velocity. Driving a car (turning its wheel) is a realistic and very simple example. A good car without driver would keep its direction. But any bump would change it and the car would go to the ditch. The driver has to correct this change. He has also to turn the wheel when the road turns. This second correction can be exerted by tonic (slow) muscular fibres. The first one has to be quick: it is exerted by phasic fibres. In a comparable way, erected station is a spontaneously unstable situation: without phasic corrections, we do not fall down at once, but when we stumble. For all these examples, artificial stability is obtained (as in some industrial devices) by inserting the mechanical system into a stabilizing loop.

3.3.4.1.2. The Components of Any Stabilizing Loop

Any stabilizing loop connects 4 components: a device posting up of the wished result (the thermostat of my heating device); some sensors measuring the actual result; a computer comparing these data to evaluate a motor order; a sufficiently powerful motor ruled by this order. In our specific case, the cerebellum is the computer; the skeletal muscles are the motors. Various sensors send data to the computer. The nature of the wished result signals is less obvious.

3.3.4.1.3. Learning

Most of the stabilized motor behaviors are not inborn, but learned or partially learned. Learning results from trial and error correcting sequences. They lead to the building of simple and fast automatisms.

3.3.4.2. How Works the Cerebellum

3.3.4.2.1. Input and Output Signals

Directly or through the relays of the brain stem, the cerebellum is connected to many components of the nervous system. It receives signals from:

-   the (tonic, slow) cortical motor areas and from the thalamus.

-   the muscles through the spinal cord. The 570 skeletal muscles are progressively recruited by 4.10^6 descending fibres (see Appendix G). The number of ascending reporting fibres is of the same order.

-   through cortical relays from the eyes, touch sensors and internal ear accelerometers (Fig. **3.18**). Note that the cerebellum of the car driver does not use all the visual data (all the pixels of the retina), but only the angle between the car and the road axis: the visual information transmitted by the visual cortex is supported by a very limited number of input cells. In the same way, the data from the accelerometers is supported by only 5.10^4 neurons [41].

These signals are weakly connected (see 1-3-2-3) to the brain stem centers. When the signals are long lasting and strong, synaptic delays and thresholds allow them to diffuse into the neighbourhood, acting on the autonomous nervous system and exciting the vomiting reflex. This is the mechanism of sea sickness.

The cerebellum sends signals toward the phasic muscular units (10^5 descending fibres) and toward the cortical motor areas..

3.3.4.2.2. Quantitative Description of the Cerebellum [42]

It contains as many neurons (40 billion) as the rest of the nervous system. But, its neurons are short. Thus, the volume of the cerebellum is only 10% of the volume of the rest of the brain. It contains several sorts of cells organized along several paths.

Microzones and One Muscle Paths: 2.10^7 "climbing cells" (the axons of the "inferior olivary nuclei", a relayin the brain stem) are connected to the same number of "Purkinje cells" which excite the "deep nuclei" in charge of the outputs. The Purkinje cells are swift transient neurons (time constant $T_D \approx 0.1$ s). Their firing rate at rest is about 30 s^{-1}, their usual frequencies in the range [0 to 100 s^{-1}] [11].

They are arranged in microzones: one Purkinje cell acts only on one muscle. There are several cells for a muscle, and all the cells associated with the same muscle are close. So, there is an association one microzone, one muscle. From the number of cells, we deduce that the inputs are the ascending and descending fibres.

The Sensorial Paths: The some 10^5 sensorial inputs are processes by a modest number of "Golgi, stellate and basket cells".

The Parallel Fibres Path: About 10^8 "mossy fibres" are connected to 4.10^{11} "granular fibres" (a surprising huge number). Their axons are called "parallel fibres". The dendrites of each Purkinje cell forms a large planar arborescence perpendicular to the parallel fibres. This geometry allows a very great number of synaptic connections (each Purkinje cell is connected to some 10^5 parallel fibres.

The sensorial paths are connected to the parallel fibres.

The mossy fibres are candidate to support the mysterious wished results signals. They are probably connected with the cortical motor areas.

3.3.4.2.3. The Cerebellum at Work

A learning period comes before the outcome of an automated one. (Learning is often called "sensor-motor calibration".)

To create a motor behavior, the nervous system (as we will see in chapter 4.2) has to define a general order (to walk, to drive the car); to define the wished result (a negligible angle between the road axis and the car); to choose the agonist and antagonist muscles and the specifications (strength, speed) of the correcting device. Most of these tasks are carried out by the cortical motor areas. The cerebellum intervenes only when the slow tonic control has been learned.

Automated Behavior (After Learning): Look at the required properties of the device governing a touch-a-target test. The global order triggers this behavior among many possible competitors. The visual input allows evaluating the distance d between the finger and the target. The now automated device has to generate a phasic excitation of the antagonist muscle. The wished strength of the phasic braking depends on the velocity v of the finger (which is the derivative of the distance). The device excites phasic descending fibres. Due to the progressive recruitment of motor units, the resulting braking force depends in an exponential way of the number of excited fibres. Efficiency requires a good agreement of the resulting braking force with the wished one.

In a weblike circuitry (for instance the cortical areas), a great number of nodes (of synaptic relays and therefore a great number of synaptic delays) allows minimizing the number of connecting lines. On the opposite, to obtain a computing time less than 0.1 second, the cerebellum has to use one synapse devices, needing a huge number of connecting lines (the granular cells).

Note also that the transmission delay between the cerebellum and the hand is relatively large (about 0.1 second): it must be corrected. In such cases, engineers use extrapolation-correction algorithms (Appendix H). Whatever the details, the main characteristic of such a device is the computation of a derivative, for instance the velocity: $v.dt = d(t) - d(t-dt)$. This implies that the signal of a neuron is delayed by another neuron and comes back to inhibit the first one. Now, such a loop (with a relay in the deep nuclei) is known: delayed signal of Purkinje cells inhibits the granular cells.

Learning: We will see in chapter 4.1 that learning is due to the activation of some post-synaptic receptors of an output neuron which is excited by a conditioning path and a conditionable one (Fig. **4.1**). The transmitter released by the conditionable path is glutamate. Now, we know what transmitter is released by each sort of cerebellum cell. Glutamate is released by climbing cells, the output neuron being the Purkinje cell. It is released by mossy fibres; the output is then a granular cell and the conditioning neuron a Golgi cell. Glutamate is also released by the deep nuclei acting on the output toward the premotor areas. These data and the discussions of chapter 4.2 (paragraph 4.2.5.5 and appendix J) are sufficient to imagine how trials and error corrections can gradually optimize a correcting device.

APPENDICES TO SECTION 3.3

A: The Huxley's model of sliding filaments

B: The Calcium and troponin cycles in muscular fibres

C: The progressive recruitment of motor units

D: Equilibrium and little perturbations of a two muscles device

E: The embedded muscular sensors

F: The descending spinal tract

G: Principle of a plausible two muscles control device

H: The cerebellar processing

Appendix A: The Huxley's Model of Sliding Filaments

Let x be the distance (in reduced unit) between a myosin head and the bound actin site) and let n(x) be the density of bound sites. The binding rate depends on some function Φ(x) and the unbinding rate on some function Ψ(x). We look at the density n(x) of bound states. The binding rate is proportional to the density 1-n of unbound states and the unbinding rate to the density n of bound states. The myosin heads move during the process. So, (as in fluid mechanics), we have to use derivative following the motion. So, the balance equation is:

$$\frac{\partial n}{\partial t} + \frac{dx}{dt}\frac{\partial n}{\partial x} = (1-n)\Phi - n\Psi$$

The term ∂n/∂t can be neglected. The velocity dx/dt is proportional to the reduced velocity u. Thus:

hu dn/dx $= (1-n)\Phi - n\Psi$

And the muscular force (due to elasticity) is proportional to $\int n(z)z\,dz$

The heart of the model is the choice of the binding unbinding functions:

Φ(x) = 1 if 0 < x < 1 Φ = 0 elsewhere

Ψ(x) = g if 0 < x < A (A > 1) Ψ very great elsewhere

Here, h, A and g are some coefficients. Thus, we obtain the differential system:

Introducing the parameters r = (1+g)/hu and ρ = g/hu, the solution is easily found:

In static conditions (u = 0): $n_{STAT} = \dfrac{1}{1+g}$ *if 0 < x < 1 and n = 0 elsewhere.*

For hu dn/dx = 1-n(1+g) if 0<x<1; hu dn/dx = -ng if 1<x<A;

n=0 if x<0 or if x>A

For contractions (u < 0):

$n = n_{STAT}\left[1 - \exp r(1-x)\right]$ *if 0 < x < 1 and n = 0 elsewhere.*

For lengthening (u > 0):

$n = n_{STAT}\left[1 - \exp(-rx)\right]$ *if 0 < x < 1*

$n = n_{STAT}\left[1 - \exp(-r)\right]\exp - \rho(x-1)$ *for 1 < x < A and n = 0 elsewhere.*

*The velocity term **Vel** (u) is the ratio of the force with a velocity u to the static force. As the force is proportional to $\int n(z).z.dz$, a simple integration gives the velocity term **Vel**. Note that this function of the variable u is continuous for u = 0, but that its derivative is discontinuous:*

*For contractions (u < 0): **Vel(u)** = 1 + (2/r²) [1 + r − eʳ]*

For lengthening (u > 0):

$$Vel(u) = 1+\left(\frac{2}{r^2}\right)\left[(1+r)\exp^{-r}-1\right]+\left(\frac{2}{\rho^2}\right)\left[1-\exp^{-r}\right]\cdot\left[1+\rho-(1+A\rho)\exp^{-\rho(A-1)}\right]$$

The parameters h and g are chosen to fit the Hill's approximation for u close to zero: g = 0.25 h = 0.75. The original paper took A = 1.5.

We have to compare this law with the very often used empirical Hill's formula [2].

$$Vel_H = \frac{1+0.25u}{1-u}$$

Numerical results for the velocity factor *vs* the lengthening velocity (in reduced unit) and for the approximate Hill velocity factor are given Table **3.8**.

Table 3.8. Hill's velocity factor.

u	4	1	0.75	0.5	0.3	0.1	0	-0.5	-1	-2	-3	-4	-10
Vel	0.7	1.57	1.73	1.86	1.9	1.57	1	0.57	0.38	0.22	0.16	0.13	0.05
Vel$_H$		∞	5.0	2.25	1.5	1.14	1	0.58	0.37	0.17	0.06	0	-0.1

Note that if the function **Vel(u)** is continuous for u = 0, its derivative is discontinuous. For weak

Appendix B: The Calcium and Troponin Cycles

The calcium-troponin-myosin cycle is summed up by the Fig. (**3.25**).

The Calcium Cycle

Each spike of a motoneuron releases some amount of the very short life acetylcholine. This transmitter opens Ca^{2+} channels in the post synaptic

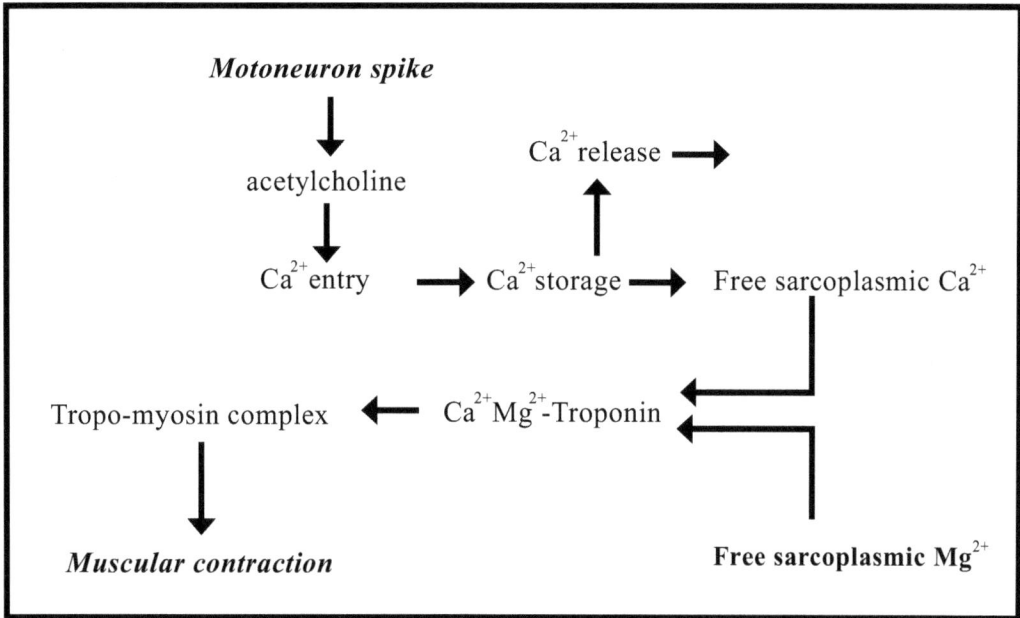

Fig. (3.25). Calcium cycle during muscular contraction.

(muscular) cell. Balance between Ca^{2+} ions entrance, storage and exit determines the sarcoplasmic free Ca^{2+} concentration. Free Ca^{2+} and free Mg^{2+} ions govern the activation of a regulating molecule, troponin C. Troponin C molecules binds to myosin to form a tropo-myosin complex in which the myosin molecule is active.

Note that one impulse exciting a motor unit induces an observable muscular twitch, that successive twitches in a motor unit are smoothed in the case of mean firing rate excitation and that the sum of the forces of several motor units smooth away any individual twitch: therefore, the muscular force behaves as a continuous function of excitation. From observation of fluorescent markers in a muscular fibre at rest and from the analysis of the summing of twitches, one find for the mean free Ca^{2+} concentration versus the motoneuron firing rate:

$[Ca^{2+}_{free}] = a + b\ (F/F_{Max})$

with a = 0.1 10^{-6} mole/L, b = 2.1 10^{-6} and F_{Max} = 60 s^{-1}

We assume that these numbers are the same for all striated muscles.

The Troponin Activation

Troponin C owns two pairs of sites S_1 and S_2 which bind to Ca^{2+}. The equilibrium constant of the reaction $Ca_{free} + S \leftrightarrow CaS$ have been measured. For S_2 sites, $K_2 = 3.2 \ 10^6 \ M^{-1}$. For sites S_1, one found $K_1 = K_2/3$. Thus, the concentrations obey to: $[CaS] = K.[S_{free}].[Ca_{free}]$ and the probability for a site S_2 to be bound is

$$\Pr ob2 = \frac{[CaS_2]}{[CaS_2]+[S_{2\,Free}]} = \frac{\xi}{1+\xi} \text{ where we have introduced the reduced variable}$$

$\xi = K_2 [Ca^2{}_{free}]$. *In the same way,* $\Pr ob1 = \dfrac{\xi}{1+3\xi}$ *(because $K_2 = 3K_1$).*

The tropo-myosin is activated if the two sites S_1 and the two sites S_2 are bound to Ca^{2+}. and all sites S_3 to Mg. Then, the ratio ρ of the actual force of a motor unit to the force for an infinite firing rate versus the actual firing rate of its motoneuron is:

$$\rho = \left(\Pr ob1\right)^2 \left(\Pr ob2\right)^2 = \left[\frac{\xi^2}{(1+\xi)(3+\xi)}\right]^2$$

Remind that the firing rate F ranges from 0 to 60 s^{-1}. From the above relations, we find the following values of ρ vs the firing rate (Table **3.9**):

Table 3.9. Actin activation *versus* the firing rate of the motoneurons.

F (s^{-1})	0	10	20	30	40	50	60
ρ	2.7 10^{-3}	0.16	0.30	0.54	0.64	0.70	0.75

Let H/H_{Max} be he ratio of the actual force (firing rate F) to the strongest force (firing rate 60). H/H_{Max} is roughly a linear function of F ranging from the minimum value $H/H_{Max} = \rho(F=0)/\rho(F=60) = \exp(-5.6) = 3.6 \ 10^{-3}$ to the maximum value $H/H_{Max} = 1$ (for F=60).

Appendix C: The Progressive Recruitment of Motor Units

The muscle is made of N motor units. All of them depend in the same way of the muscular length and velocity, but have different number of fibres and receive different excitations. In this appendix, we look only at the static forces for optimum length. Let F_k be the actual force of the k motor unit (for static conditions and optimum length).

The total force of the muscle is $F_{TOT} = F_1 + F_2 + ... + F_N$, which can Let Φ_k be the maximum force of the k motor unit. They can be numbered according their size (their number of fibres), starting from the weakest: $\Phi_1 < \Phi_2 < ... < \Phi_N$.

Experiments show that the force Φ_k is roughly an exponential function of the index k. *We can interpret this property as resulting of a functional optimization: the best functional choice of the maximum forces Φ_k is such (for a tonic muscle) that the accuracy of the position does not depend on the load. A variation of the control potential δU of the control potential of neuron k would have to cause a variation δF_{TOT} proportional to F_{TOT}.*

We neglect here the residual force factor R_0. If the motor neuron has a linear input-output law: $\delta F = \Psi_D \, \delta U$ where Ψ_D is some constant, then $\delta F_{TOT} = [(\Psi_D / F_{Max}) \delta U] \, \Phi_k$ The bracket does not depend on k and is obviously a constant in k. Thus:

$\delta F_{TOT} = \alpha \Phi_k$ *and* $d \, \delta F_{TOT}/dk = \alpha \, d\Phi_k/dk$

If N was infinite, F_{TOT} would be an integral:

$$F_{TOT}(k) = \int_1^k \Phi(x)dx \ or \ dF_{TOT}/dk = \Phi_k$$

Then, the wanted condition δF_{TOT} proportional to F_{TOT} can be written $d\delta F_{TOT}/dk$ proportional to dF_{TOT}/dk, leading to

$\alpha \, d\Phi_k/dk = \Phi_k$

where α is some unknown constant. The solution is an exponential. For an equation with discrete values of k, $\Phi_K = \Phi_1 e^{\alpha(K-1)}$ is the best approximate solution.

First Approximation

In a first approach, we assume all or none excitations of the motor units:

$F_k = 0$ or $F_k = \Phi_k$.

We have then to introduce a <u>recruitment law</u> (probably due to some neuronal connections in the spinal node): weak motor units are excited before the strongest ones.

$F_k = \Phi_k$ *if k is in the range [k=1, k=K] and* $F_k = 0$ *if k > K*

(Thus, the configuration $F_1 = \Phi_1$; $F_2 = \Phi_2$; $F_3 = 0$ *is allowed but* $F_1 = \Phi_1$; $F_2 = 0$; $F_3 = \Phi_3$ *does not occur)*

Then: $F_{TOT}(K) = \Phi_1 + \Phi_2 + ... + \Phi_K$

From the exponential law, we deduce easily: $F_{TOT}(K) = \Phi_1 \left(\dfrac{e^{\alpha K} - 1}{e^{\alpha} - 1} \right)$

Continuous Case

The firing rate of real motoneurons ranges from 0 to 60 s^{-1}. Then, the force of the k motor unit is $F_k = R.\Phi_k$ (R, a function of the firing rate, was defined in Appendix 13; for the present time, we neglect R_0). The recruitment law becomes:

$F_k = \Phi_k$ if k < K $F_K = R.\Phi_K$ $F_k = 0$ if k > K

Using the same exponential law as above, we find that the total force can take any value between 0 (K=1, R=0) and $F_{TOT}(N) = \Phi_1 \left(\dfrac{e^{\alpha N} - 1}{e^{\alpha} - 1} \right)$ with K=N, R=1.

Remark: Let G be a number, the integer part of which is N and the non integer R. Then $F_{TOT}(G)$ *is a function of G made of a continuous sequence of linear segments.*

Orders of Magnitude

Look at the biceps femoris (see III.5.2): it owns N = 8 tonic motor units. The range $U_{MAX} - U_{Thr}$ *of the control potential of the exciting neurons is 24 mV with a 0.15 mV accuracy due to the noise. Therefore the number of possible messages that an exciting neuron can transmit is 24/0.15 =160. We compute* $F_{TOT}(N)/\Phi_1$ *versus α and the minimum exerted force* $\Phi_1/160$ *for a maximum force* $F_{TOT}(N) = 100$ kg:

$\alpha = 0.1$ $F_{TOT}(N)/\Phi_1 = 12$ $F_{MIN} = 0.5$ kg

$\alpha = 0.2$ $\boldsymbol{F_{TOT}}(N)/\boldsymbol{\Phi_1} = 18$ $F_{MIN} = 0.3$ kg

Effect of the Minimum Force Factor R_0

This effect is negligible except for weak exerted forces.

The recruitment law is now: $\boldsymbol{F_k} = \boldsymbol{\Phi_k}$ if $k < K$ $\boldsymbol{F_K} = \boldsymbol{R}.\boldsymbol{\Phi_K}$ $\boldsymbol{F_k} = \boldsymbol{R_0}$ if $k > K$

We find then for K integer in the range [1, N] and with R in the range [R_0, 1]:

$$F_{TOT}(K) = \Phi_1\left(R_0 \frac{e^{\alpha N} - 1}{e^{\alpha} - 1} + (R - R_0)e^{\alpha K} + (1 - R_0)\frac{e^{\alpha(K-1)} - 1}{e^{\alpha} - 1} \right)$$

For N = 8 and α = 0.1 or 0.2, we obtain $F_{TOT}(1)=1.038\Phi_1$ or $1.061\Phi_1$ when the minimum exerted force is $\Phi_1/160 = 0.006\Phi_1$. The effect is negligible if K > 1 or if K =1 and R-R_0 >0.1.

Appendix D: Equilibrium and Little Perturbations of a Two Muscles Device

The master equation is:

$ML_0\, d^2y/dt^2 = Mg +$ antagonist force $-$ agonist force

Reduced Units

We introduce

a)   *the reduced lengths $y=L_{AGONIST}/L_0$ and the similar z for the antagonist. They are linked by: $y + z = 1.6$ (and by $dy/dt = -dz/dt$).*

b)   *the reduced mass $\mu = Mg/F_{.MAX}$ where F_{MAX} is the total maximum force for each of the muscles.*

c)   *the characteristic time $\theta = 0.1$ s and the pendulum time $\tau=\sqrt{L_0/g} = 0.2$ s (with $L_0 = 40$ cm).*

d)   *the length factors of the agonist muscle $\alpha = $ exp-[6.25(1-y)^2]_and of the antagonist muscle $\beta =$ exp-[6.25(1-z)^2)] = exp-[6.26(0.6-y)^2].*

e) *the reduced velocities: In the sliding filaments velocity term, we use the u values* $u_1 = \theta \dfrac{dy}{dt}$ *for the first muscle and* $u_2 = -\theta \dfrac{dy}{dt}$ *for the antagonist muscle.*

f) *the recruitment factors Φ for the agonist and Ψ for the antagonist.*

Table 3.10 gives the two length factors as a function of the reduced length.

Table 3.10. The reduced length factors.

y	1.05	1.0	.95	.85	.75	.65	.60
α	.98	1.0	.98	.87	.68	.47	.37
β	.28	.37	.47	.68	.87	.98	1.0

The Master Equation

With these reduced units, the motion obeys to:

$$\mu\tau^2 = \mu - \alpha\Phi Vel(u1=\theta dy/dt) + \beta\Psi Vel(u2=-\theta dy/dt)$$

The Equilibrium Condition

In a static position: $d^2y/dt^2 = dy/dt = 0$. Then Vel = 1. Phasic excitations are transient; so, at equilibrium, we have to deal only with the tonic excitations. The equilibrium equation states that the load is equal to the difference between the two muscular forces:

$\mu = \Phi\alpha - \Psi\beta$.

Defining the Tonus at Equilibrium

Tonus is generally and imperfectly defined as the force exerted by a muscle at rest. We have to apply this intuitive notion to a two muscles device.

We call **Tonus at equilibrium T** the arithmetical average of the agonist and antagonist reduced forces at equilibrium: T = 0.5($\Phi\alpha + \Psi\beta$).

A Linearized Master Equation

For little perturbations, we have:

$$y = y_{Equ} + \xi(t) \quad \alpha = \alpha_{Equ}\left[1 + 12.5\left(1 - y_{Equ}\right)\xi\right] \quad \beta = \beta_{Equ}\left[1 - 12.5\left(y_{Equ} - 0.6\right)\xi\right]$$

$$Vel = 1 + 1.25u \ \ if \ u = \theta\frac{d\xi}{dt} < 0 \ (contracting)$$

and $Vel = 1 + 6.17u \ \ if \ u > 0$ *(lengthening)*

Searching solutions of the first order equation behaving as exp (-st), we find:

For contraction of the first muscle (dy/dt < 0):

$$\mu\tau^2 s^2 - 7.42T\theta s + 2.46\mu\vartheta s + 12.5\left[\left(0.8 - y_{Equ}\right)\mu + 0.4T\right] = 0$$

And for lengthening of the first muscle (dy/dt>0):

$$\mu\tau^2 s^2 - 7.42T\theta s - 2.46\mu\vartheta s + 12.5\left[\left(0.8 - y_{Equ}\right)\mu + 0.4T\right] = 0.$$

In our example, θ = 0.1 s and τ = 0.2 s. From these equations, we evaluate s for any set of the 3 parameters y_{Equ}, T, μ. We find either 2 damping coefficients f_1 and f_2 or a complex value $f \pm i\omega$. Note that, if y<1, the damping coefficients f are >0.

Appendix E: The Embedded Muscular Sensors

Type II spindles

The Response of Agonist Spindles: The spindles are excited by mechano sensors giving a component of the control potential proportional to y. They are inhibited by γ tonic motoneurons, which are standard slow neurons with firing rate $F_{\gamma \ tonic}$ in the range [0, 60 s^{-1}]. The spindle seems to be a transient swift neuron (threshold 3.3 mV, $F_D \approx 10$ s^{-1}). *We assume that the control potential U of the spindle is a linear function of the inputs: U = a + by -c $F_{\gamma \ tonic}$ and that the output firing rate is a linear function of the control potential. To determine the coefficients, we assume that the device is optimised. As seen in part I, the inhibiting potential cannot* *exceed – 10 mV. Thus, the inhibiting term would be -10 $(F_{\gamma\,tonic}/60)$. An optimal condition is*

$U = U_{Thr}$ for $y_{min} = 0.65$, $F_{\gamma\,tonic} = 0$ or for $y = y_{max} = 0.95$ and $F_{\gamma\,tonic} = 60$. Then:

$$U = U_{Thr} + 10\left[\frac{(y - y_{min})}{(y_{max} - y_{min})} - \frac{\gamma_{TONIC}Firing \cdot rate}{Maximum \cdot Firing \cdot rate}\right]$$

As the maximum firing rate of the fibre II is 60 s^{-1}, we find:

$FiberII \cdot firing \cdot rate = 200(y - y_{wanted})$ or 0 if $y < y_{wanted}$

with $y_{wanted} = 0.65 + \dfrac{\gamma_{TONIC}\,firing \cdot rate}{200}$.

The measurement of the length y:

The two muscles device owns agonist and antagonist spindles. At first view, the two type II spindles would furnish two signals (the frequencies of fibres II and of fibres II^A) when we need only one. But we remark that the fibres II firing rate is proportional to y-y_{wanted} whenever $y > y_{wanted}$ and are mute whenever $y < y_{wanted}$ while fibres II^A are firing when fibre II are mute (because fibres II^A measure $z = 1.6$-y). Thus, the set of the two type II spindles furnishes in fact only one significant signal: the measurement of $(y - y_{Wanted})$.

Golgi's Tendon Organs

The Response of Agonist Spindles: They are excited by β motoneurons (range of the firing rate: 0, 60 s^{-1}) and inhibited by a mechanosensor, the opening of which is proportional to the muscular force. The maximum firing rate of the output (the fibre Ib) is 90 s^{-1}. Reasoning as for the type II spindles, we find:

$$FiberIb \cdot firing \cdot rate = 90\left[\frac{Force}{ForceMax} - \frac{Motoneuron\beta \cdot firing \cdot rate}{60}\right]$$

or 0 if $U < U_{Thr}$

Measurement of the Load: The 2 outputs (frequencies of fibres Ib and of fibres Ib^A) depends on 4 inputs (forces F and F^A, frequencies of the motoneurons β and

β^A). The wiring has to extract 1 order from this set. A trick is to use only one kind of motoneurons β: β firing rate $= \beta^A$ firing rate. Then the difference of Ib and Ib^A firing rates measures the kinetic load $\mu_{Kin} = (F-F^A)/F_{Max}$, which is equal to μ when the equilibrium is reached. Note that a progressive recruitment of output neurons (excited by the Ib and Ib^A) fibres allows to measure $Log(\mu_{KIN})$, which is never far from $Log(\mu)$.

Type I spindles

Description: The total length y (assigned by the muscle) is the sum of the lengths of the two parts: $y = x + z$ (Fig. **3.26**).

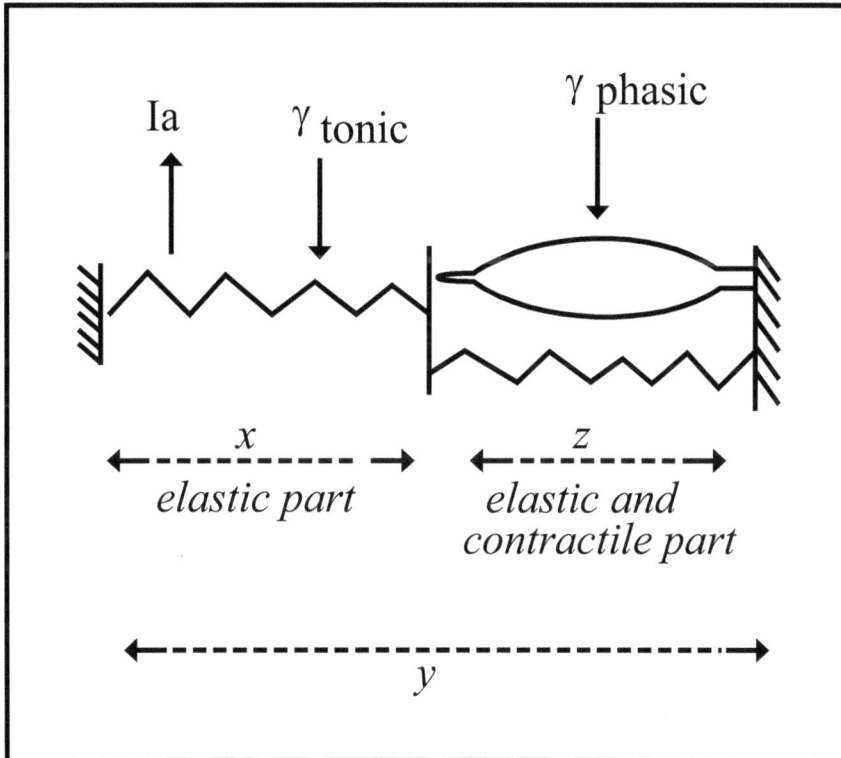

Fig. (3.26). Scheme of Type I spindle.

Now, the forces of the two parts have to be equal. When the firing rate of the γ phasic motoneurons increases, the second part contracts itself; z decreases, hence x increases. The first part of the cell is very similar to a neuron cell body. Its axon is the Ia fibre. The cell body is excited by the opening of ionic channels due to the strain, (thus its control potential U is propotional to x) and is inhibited by some γ tonic motoneurons.

This neuron like cell is a swift neuron with an adaptation time great when compared to the motion duration) and with an initial all or none response with high firing rate (about 500 s^{-1}) as soon as the control potential U is greater than the threshold.

In a rough approximation, fibres Ia generate a short burst of high firing rate spikes if y − y_{WANTED} is greater than some threshold and the γ phasic firing control the value of this threshold.

A Possible Detailed Model

Length x of the Active Segment: *The lengths of the two components are* x *and* z *with* y = x + z. *Let* φ_X *and* φ_Z *be the forces of the two components. At each time, we must have:* $\varphi_X = \varphi_Z$ *The first component is purely elastic:*

$\varphi_X = \Phi_X (x − x_1)$ *where* Φ_X *and* x_1 *are two constants. The second force is the sum of an elastic and a contractile term:* $\varphi_Z = \Phi_Z (z − z_1) + \Phi_C f \exp- [(z-z_{MAX})/0.4z_{MAX}]^2$

where Φ_Z Φ_C *and* z_1 *are three constants and where* f *is the* γ *phasic firing rate divided by the maximum firing rate (60 s^{-1}).*

For simplicity sake, we assume that x_1 = z_1 *and that* Φ_X = Φ_Z = 1 *in some reduced unit. Then, the equations of the system are* y = x + z *and* (x − x_1) = (z − x_1) + $\Phi_C f \exp- [(z-z_{MAX})/0.4z_{MAX}]^2$

As usual, the muscle length y = L/L_0 *is in the range [0.65,0.95]. In the same way, the range of the contractile element is 0.95 > z/z_{MAX} > 0.65.*

Nul force occurs when f = 0, x = z = x_1. *Now, we expect a nul force for the minimum muscle length* y = 0.65. *Then,* x_1 = 0.65/2 = 0.325.

The maximum value of z *occurs when* y *is maximum and when the contractile force is nul. Then,* y = 0.95 *and (because* f = 0) x = z. *Therefore,* z = 0.95/2. *Now, this z value is the maximum of the z range, which is 0.95 z_{MAX}. Hence, z_{MAX} = 0.5 and the contractile term can be written:* $\Phi_C f \exp- [5z - 2.5]^2$. *The equations* y = x + z *and* $\varphi_X = \varphi_Z$ *become:*

$x = z + \Phi_C f \exp\text{-}(5z\text{-}2.5)^2$ *and* $y = 2z + \Phi_C f \exp\text{-}(5z\text{-}2.5)^2$

A good approximation of x vs. f and y is:

x = 0.5 [y + 0.15f + 0.2fy] with **f = γphasic firing rate/60**

Ia firing rate as a function of x

The strain and the entering ionic current are proportional to $x - x_1$. Thus:

U *(mV)* **= b(x – 0.325) -c.γtonic firing rate**

where b and c are some constants. Now, the active segment behaves as an adaptative swift neuron with such a great time constant ($T_D \approx 0.3$ s) and a threshold U_{Thr} = 3.3 mV. The firing rate F_0 is well fitted by $F_0 = 250 (U – 3.3)^{0.2}$.

Appendix F: Principle of a Plausible Two Muscles Control Device

It furnishes the muscular length y(t) after a brisk change Δy_{WANTED} of the order.

The System Obeys to the Master Equation

$x = \theta dy/dt$ *and:* $\mu\tau^2 dx/dt = \mu - \alpha(\Phi + \Phi_P) Vel(u1 = x) + \beta(\Psi + \Psi_P) Vel(u2 = -x)$

The muscle time constants are θ = 0.1 s and τ = 0.2 s in our example. The P index characterizes the phasic excitations and we assume that the maximum forces of the tonic and of the phasic components are equal. The excitations Φ and Ψ are exponential functions of the numbers k and k^A of recruited tonic motor units:

Φ **= exp(5.6-0.7k)** and: Ψ **= exp(5.6-0.7k^A)**

An Asymmetric Control

From the properties of the imbedded sensors, we associate the Golgi's sensors to the control of the adaptation to the load and the type II spindles to the control of length. From the unexpected properties of muscles, we associate the agonist muscle to the load control and the antagonist muscle to the length control. At last, we assume that the linearly controlled parameters are k and k^A (and not their exponentials Φ and Ψ).

The Effective Load

The first idea is to use a measured value of the load. But the Golgi sensors do not measure the load μ; they measure the kinetic load $\mu_{Kin} = \mu\left(1 - \tau^2 \dfrac{d^2 y}{dt^2}\right) = \dfrac{F - F^A}{F_{Max}}$.

Using a slow synapse, a neuron compute then the convolution:

$$\mu_{EFFECTIVE} = \frac{1}{T}\int \exp\left(\frac{\theta - t}{T}\right)\mu_{Kin}(\theta)\,d\theta$$

And we assume that the control loop uses $\mu_{EFFECTIVE}$ in place of μ.

The Best Excitation and k Value

We have seen that the excitation $\Phi = 2\mu/\alpha$ law leads to $1/f \approx \theta$ if $\mu < 0$. So, we choose $\Phi_{WANTED} = 2\mu_{EFFECTIVE}/\alpha$ if $\mu < 0.5\alpha$, else $\Phi = 1$. Noting that $Log(\Phi)$ is a linear function of k and that progressive recruitment allows to measure $Log(\mu)$, we introduce k_{WANTED} proportional to $Log(\Phi_{WANTED})$.

Two Control Loops

A simple delayed loop allows controlling k: $\dfrac{dk}{dt} = \left(\dfrac{1}{T_1}\right)\left(k_{WANTED} - k\right)$

Another loop controls the antagonist k^A to obtain the wanted length $y_{WANTED:}$:

$\dfrac{dk^A}{dt} = \left(\dfrac{1}{T_2}\right)\left(y_{wanted} - y\right)$

The Phasic Orders

A rough approximation is: $\Phi_P = z * \left[1 - \exp\left(\dfrac{\Delta y + y_{WANTED} - y}{0.5\Delta y}\right)\right]$ *if $y > y_{WANTED} + \Delta y$*

or $\Phi_P = 0$ else. Δy is a constant and z an evaluation of the Renshaw's inhibition):

$\dfrac{dz}{dt} = \dfrac{1-z}{T_P}$ *if the phasic neuron is mute or* $\dfrac{dz}{dt} = \dfrac{-z}{T_P}$ *if it is firing.*

The same equations apply to the antagonist phasic neuron.

Parameters Used in our Examples

μ-convolution time T = 0.2 s (slightly greater than the characteristic time θ). Time constants of the control loops: $T_1 = T_2 = 0.01$ s (an order of magnitude greater than the spike duration).

Renshaw's neuron time constant $T_P = 0.03$ s (close to the experimental value). Threshold of the phasic action: $\Delta y = 0.005$ (coherent with the y range).

Appendix G: The Spinal Tracts

The Descending Pathways

The pyramidal tract goes through the spinal cord from the brain to the muscles. The tract is composed of $4 \cdot 10^4$ fast fibres (each one linked to one phasic motor unit) and of $2 \cdot 10^6$ slow fibres (about 10 fibres for one tonic motor unit). Fast fibres excite probably the phasic γ motoneurons. Slow fibres excite β and tonic γ motoneurons, excite or inhibit Renshaw neurons (see 3.3.3.5) and (at the highest spinal hierarchical level) modulating the spinal automatisms. They probably do not excite directly any α motoneuron. But remark that the face and the hands are governed by fibres escaping at a very high level from the spine. Thus, the total numbers of phasic exciting fibres is of the order of 10^5 and the number of tonic ones about 5.10^6.

The descending fibres are distributed among two groups, each of them separated into a direct and an indirect pathway. The first group, located in the central part of the medulla, governs the spinal automatisms and the simplest movements of arms and legs. Its indirect pathway makes synapse in the reticular formation of the brain stem. The other group (which appears only in evolved mammals) is located in the lateral part of the medulla. It has in charge the fine individual command of muscles. It plays a main part in the precise command of the hands and of the head (in particular during phonation). Its indirect pathway makes synapse in the red nucleus of the brain stem.

The Ascending Spinal Tract

In the leg, a first neuron goes from each of the above sensors to the medulla. A second neuron goes from the medulla to the thalamus. Starting from synapses with the same first neuron, other neurons go up and down inside the spinal cord, in which they act as interconnecting neurons. After a thalamic relay, all the pathways reach the cortex. They excite the inputs of several columnar sub-units. Close sensors excite close columns; thus, each sub-unit is a map of the body. The size of the cortical device devoted to a part of the body is proportional to the number of neurons coming from this part (to the number of sensors): the projection of the hands is much larger that the projections of the forearms.

Appendix H: The Cerebellar Processing

We look at the case of touch-a-target test. Visual signals allow evaluating the target distance x(t) at time t. In such cases, engineers use extrapolation-correction algorithms. Let δt be the delay of the feedback inhibiting loop. The algorithm compute forecasted values from the observed one.

If:

$\hat{x}(t)$ is the forecasted value of x at time t,

$\underline{x}(t)$ is the estimate of x at time t,

$x(t)$ is the observed value of x at time t

v(t) = x(t) –x (t- δt)

e(t) = (observed – forecasted values) is the error

We use the following chain of equations:

$$\underline{x}(t) = \hat{x}(t) + K\left(x(t) - \hat{x}(t)\right) = \hat{x}(t) + K\varepsilon$$
$$\hat{x}(t + \delta t) = \underline{x}(t) + v\delta t$$

Resulting of the choice made for the value of the gain K, (included between 0 and 1), more weight is given either to the forecasted value or to the observed value. If K=1 the estimate is equal to the observed value, there is no filtering; if K = 0, there is no correction, the estimate is equal to the forecast and the observed value is rejected.

The purpose of such algorithms is to reduce the error incurred on each observation by combining a time sequence of several observations. It is well known that the best estimate of a fixed value is an average of N measurements since the error due to the noise is reduced by $\sqrt{N}$. This approach is impossible to apply to values that change with time. If the equations driving the changes are more or less known [in our very simple example the equation of changes is a brute equation of motion: $\mathrm{x}\left(t+\delta t\right)=x(t)+v.\delta t$ *where v is supposed to be known], sequential algorithms can be used to compute the best estimate (in a well defined sense) of the value at the date t of the last measurement (filtering process) or at the future date t+n.δt (extrapolation filter). The speed v and the acceleration can be fitted as well in the same recurrent process.*

REFERENCES

[1]   Buck LB. Receptor diversity and spatial patterning in the mammalian olfactory system. Ciba Found Symp 1993; 179: 51-64. Review.

[2]   Mori K, Nagao H, Yoshihara Y. The olfactory bulb : coding and processing of odor molecule information. Science. 1999; 286(5440): 711-5. Review.

[3]   Coile MC. Visual apparatus of dog. Dog World. 1997 December; pp. 44-9.

[4]   Hartline H.K, Ratliff F. Spatial summation of inhibitory influences in the eye of Limulus and the mutual interactions of receptors units. J Gen Physiol 1958; 41: 1049-66.

[5]   Granit R. Neural activity in the retina. Chapter XXIX in Handbook of Physiology-Neurology I. Field J, Magoun HW, Hall VE ed. Am Physiological Society, Washington DC 1959: pp 693-712.

[6]   Kuffler SW. Neurons in the retina: organization, inhibition and excitation problems. Cold Spr Harb Symp Quant Biol 1952; 17: 281-92.

[7]   Chaffee EL, Sutcliff E. Elelectroretinogram of the cold blooded horned toad. Am J Physiol 1930; 95: 250-4.

[8]   Weng S, Sun W, He S. Identification of ON-OFF direction-selective ganglion cells in the mouse retina. J Physiol 2005; 562(Pt 3): 915-23.

[9]   Morris JS, deBonis M, Dolan RJ. Human amygdale responses to fearful eyes. Neuroimage 2002; 17(1): 214-22.

[10]   Hubel DH, Wiesel TN. Receptive fields and functional architecture in two nonstriate visual areas (18 and 19) of the cat. J Neurophysiol 1965; 28: 229-89.

[11]   Llinas R, Sugimori M. Electrophysiological properties of *in vitro* Purkinge cell dendrites in mammalian cerebellar slices. J Physiol 1980; 305: 197-213.

[12]   Nieder A, Miller EK. Coding of cognitive magnitude: compressed scaling of numerical information in the primate prefrontal cortex. Neuron 2003; 37(1): 149-57.

[13]   Tranel D, Damasio AR, Damasio H. Intact recognition of facial expression, gender, and age in patients with impaired recognition of face identity. Neurology 1988; 38(5): 690-6.

[14]   Adolphs R, Damasio H, Tranel D, Damasio AR. Cortcal system for the recognition of emotion in facial expressions. J Neurosci 1996; 16(23): 7678-87.

[15]   Whalen PJ, Kagan J, Cook RG, *et al*. Human amygdale responsivity to masked fearful eye whites. Science 2004; 306(5704): 2061.

[16]   Hillerman T. The Dance Hall of the death. Harper Collins Ed., New York 1973. pp 117-8.

[17]   Békésy G von. Some Biophysical experiments from Fifty Years Ago. Ann Rev Physiol 1974; 36: 1-16.

[18]   Altman JA, Kalmykova IV. Role of the dog's auditory cortex in discrimination of sound signals simulating sound source movement. Hear Res 1986; 24(3): 243-53.

[19]   Brosch M, Schreiner CE. Sequence sensitivity of neurons in cat primary auditory cortex. Cereb Cortex 2000; 10(12): 1155-67.

[20]   Knudsen EI, Konishi M. A neural map of auditory space in the owl. Science 1978: 200(4343): 795-7.

[21]   Middlebrooks JC, Knudsen EI. A neural code for auditory space in the cat's superior colliculus. J Neurosci 1984; 4(10): 2621-34.

[22]   Lorenz K. Das sogenannte Böse. Zur Naturgeschichte der Aggression. Borotha-Schoeler Ed. Wien, 1963; pp. 294.

[23]   Fowler CJ, Griffiths D, de Groat WC. The neural control of micturition. Nat Rev Neurosci 2008; 9(6): 453-66.

[24]   Olds J, Milner P. Positive reinforcement produced by electrical stimulation of septal area and other regions of rat brain. J Compar Physiol Psychol 1954; 47: 419-27.

[25]   Lorenz K, Ed. Studies on Animal and Human Behavior. Vol II. London, Methuen Ed.,1971, p. 327.

[26]   Huxley A. Muscle structure and theories of contraction. Progress Biophys Biophys Chem 1957; 7, 255-318.

[27]   Hill A. V. The mode of acrion of nicotine and curare determined by the form of the contraction curve and the method of temperature coefficients. The Journal of Physiology. 1909; 39(5): 361-73.

[28]   Valentini F, Nelson P. The force-excitation relation in a striated skeletal muscle fibre. C R Acad Sc III. 1984; 298(19): 545-8. French.

[29]   Matthew B. Nerve endings in mammalian muscles. *J. Physiol.* 1933; 78: 1-53

[30]   Houk J, Simon W. Responses of Golgi tendon organs to forces applied to muscle tendon. J Neurophysiol. 1967; 30(6): 1466-81.

[31]   Laporte Y, Emonet-Dénand F. Neromuscular spindles. Arch Ital Biol 1973; 111(3-4): 372-86. Review. French.

[32]   Hasan Z. A model of spindle afferent response to muscle stretch. J Neurophysiol 1983; 49(4): 989-1006.

[33]   Prochazka A. Quantifying proprioception. Prog Brain Res 1999 ; 123 : 133-42. Review.

[34]   de Schotten MT, Urbanski M, Duffau H, *et al.* Direct evidence for a parietal-frontal subserving spatial awareness in humans. Science 2005; 309: 2226-8.

[35]   O' Keefe J, Dostrovsky J. The hippocampus as a spatial map. Preliminary evidence from unit activity in the freely-moving rat. Brain Res 1971; 34(1): 171-5.

[36]   Ekstrom AD, Kahana MJ, Caplan JB, *et al.* Cellular networks underlying human spatial navigation. Nature 2003; 425(6954): 184-8.

[37]   Todorov E. direct cortical control of muscle activation in voluntary arm movements: a model. Nat Neurosci 2000; 3(4): 391-8.

[38]   Scherrington C. Integrative Action of the Nervous system, Cambridge Univ. Press, London, 1947.

[39]   Moruzzi G, Magoun H.W. Brain stem reticular formation and activation of the EEG. Encephalogr Clin Neurophysiol 1949; 1(4): 455-73.

[40]   Wachholder K, Altenburger H. Beitrage zur Physiologie der willkürlichen Bewegungen. Pflügger Arch 1925; 209: 286-300.

[41]   Ozdogmus Ö, *et al.* Connections between the facial, vestibular and cochlear nerve bundles into the internal auditory canal. J Anat 2014; 205(1): 65-75.

[42]   Apps R, Garwicz M. Anatomical and physiological foundations of cerebellar information processing. Nat Rev Neurosci 2005; 6(4): 297-311.

CHAPTER 4

Learning and Memory

Abstract: Learning occurs when excitations of a neuron change its excitability. At the cell level, some effects are observed just after birth. From then two phenomena occur: the first, which governs for instance sleep duration, is the adaptation of the neuron metabolism to its needs (learning delay: one month; instantaneous reading; forgetting delay: one month). The other, which explains pavlovian linking, is supported by special post-synaptic receptors (learning delay: 10 minutes; instantaneous reading; forgetting delay, several years). At the system level, recording needs the simultaneous excitation of a conditional, a conditioning and an emotional (limbic) modulating pathway. Circuitry of recognizing flowers device in honeybee is very simple. In Pavlov's dog, learning has to link a great number of possible conditional pathways to a great number of possible conditioning ones (a problem solved by the building of the web to link a great number of phones): the switching circuitry uses many elementary Pavlov's link. Disappointment (a modulating signal) causes differentiation. Sequential links allow memorizing a melody or a sequence of words. If supplemented with a backward transmission, such circuitry generates recollection, building up of abstract concepts and partial recollection of never transmitted signals, known as imagination. Links between real or virtual sensorial inputs and motor orders play a main role. The abstract "me personally" is an abstract concept resulting from links between motor orders and report of an environmental change.

Keywords: Backward signals, concepts, conditional pathway, differentiation, forgetting, glutamate receptors, honeybee, imagination, learned motions, learning, long-term memory, moto-sensorial devices, neo natal plasticity, pavlovian link, Pavlov's dog, pleasure impulse, reading of memory, recollection, self-referent regulations, sleep duration.

4.1. Memory at the Cell Level

4.1.1. The Pavlov's Reflex Link

Main fact: glutamate NMDA synapses support usual memory.

Introduction: We have to show (Section 4.2) that all kinds of long-term memory, learning and conditioned reflexes are complex macroscopic phenomena built around an elementary circuitry, the simplest example of which is the elementary Pavlov's link.

Elementary links and conditioned reflexes have been observed in many animals, from bees till vertebrates. In man, most of the parts of the CNS (temporal cortex, amygdalae and limbic system, hippocampus, diencephalon, cerebellum, pre-motor cortex…) seem to play a part in the memorizing process.

4.1.1.1. Swift Glutamate Receptors

Glutamate is the most usual exciting transmitter of the CNS. It is present in over 50% of nervous tissue. It acts on several kinds of receptors: slow ones acting through a G protein (official name: metabotropic (mGluR) receptors) and swift receptors acting directly on ionic post synaptic channels. Their official name refers to the (artificial) molecules which bind with them. So, one distinguishes NMDA, AMPA, kainates receptors. Memorizing is only linked to NMDA ones (NMDA is an abbreviation for N-Methyl-D-Aspartate). We will describe hereafter their properties.

Experiments reveal a relation between NMDA receptors and learning ability.

On the contrary of normal mice, those with blocked NMDA genes are unable to learn their way in a maze [1, 2]. Another genetic manipulation led to "doogie" mice (with increased NMDA concentration). They learned faster and remember more than normal ones [3].

4.1.1.1.1. The Simplest Circuitry

The simplest workable circuitry is composed of three neurons: a conditioning one and a conditionable (or conditional) one. Excitations from various sensorial or emotional sources combine to make them fire with firing rates F_1 and F_2. They act as the inputs of an output neurone (Fig. **4.1**), which is an aperiodic neuron. The synaptic couplings are weak. Then, the response of the output is proportional to the input excitations.

4.1.1.2. The Post-Synaptic Glutamate Receptors

The post-synaptic membrane of every swift glutamate synapse contains a mixture of two kinds of swift receptors: *main receptors* (called AMPA/kaïnates) and *activating receptors* (called NMDA).

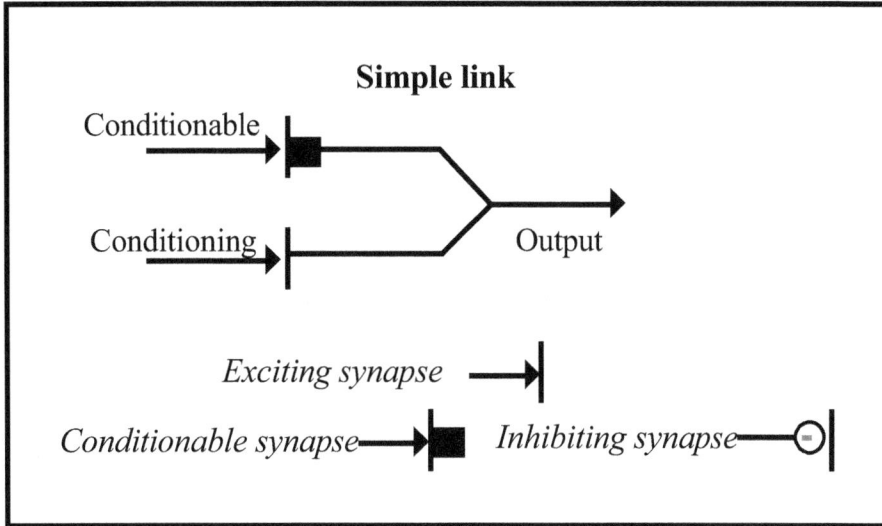

Fig. (4.1). The simple Pavlov's link.

4.1.1.2.1. The Main Receptors

They can be in operational or in non operational state: when bound to a transmitter (glutamate) molecule, the operational receptors (and only them) open exciting post-synaptic channels. Let n_R be the proportion of operational receptors. In the initial state, $n_R = 0$. Learning is an increase of the number n_R of operational main receptors.

4.1.1.2.2. The Activating NMDA Receptors

Their role is to increase the ratio n_R. The associated channels own a double gating (see Appendix A). One gate is open when the NMDA receptor is bound to a glutamate molecule, the other one when the cell body potential is greater than 0 mV, which is during a spike. Thus: if the excitation of the other synapses triggers a spike and if, simultaneously, the NMDA receptor is excited by glutamate, then the NMDA channel opens and a small amount of Ca^{2+} enters into the cell [4].

These Ca^{2+} ions act as a messengers: they induce the synthesis of an activating protein which is transported toward the main receptors and transforms non operational receptors into operational ones: n_R increases. In man, the duration of this cycle is about 10 minutes. This delay is probably shorter in small animals such as honeybees.

4.1.1.3. Recording, Reading, Forgetting

A memorizing device using a magnetic tape is able to record, read and delete messages. We look at the responses of the system shown in Fig. (**4.1**). It is able to record and to read. But progressive forgetting takes the place of deleting.

The control potential potential of the output neuron depends on the firing rates F_1 and F_2 of the input neurons and on the actual value of n_R (Appendix C). Continuous firing leads 10 minutes later to a progressive increase of n_R (Appendic B). Combining these two schemes (Appendix C), we describe recording and reading: the simultaneous firing of both the conditioning and the conditional neurons has created an almost permanent link between the output and the conditional input: it is the mechanism assumed by Pavlov.

4.1.1.3.1. Memory Losses

On the contrary of computer memories, the circuitry of Fig. (**4.1**) does not include any deleting device. But there is a spontaneous decrease of n_R. Results from psychological tests lead to set up a law that relates n_R to time with a very great slope near the origin of time and a very weak slope after a long delay. Repulsive forces between operational receptors (Appendix D) can explain such a law, leading to the Table **4.1**.

Note that, with such a law, old remembrances are masked, but are not cancelled.

Table 4.1. Progressive decrease of the number of functional receptors.

n_R	1	.9	.8	.7	.6	.5	.4	.3	.2	.1
t	0	4 h	20 h	4 days	16 d	2.3 m	10 m	4 year	20 y	130 y

Self Regeneration: After each simultaneous firing of the output and of the conditional input, n_R increases. Links that are often used never degenerate. (A possible function of dreams could be to reinforce the links partly deactivated).

4.1.1.4. Useful Variants

We look at simple devices with only one glutamate swift synapse.

4.1.1.4.1. The Response of the Simplest Circuitry

The system of Fig. (**4.1**) creates a link between the conditional neuron and the old output associated to the conditioning neuron. We summarize its functioning by looking at the firing of the output when the conditioning neuron alone or the conditional neuron alone or both are firing before or after recording:

Before learning:

    Conditioning neuron alone → output neuron fires

    Conditional neuron alone   (does not fire)

    Both neurons   → output fires, delayed n_R increase

After learning

    Conditioning neuron alone→ output neuron fires

    Conditional neuron alone   → output neuron fires

This device acts qualitatively as a diode memory.

4.1.1.4.2. Opening a New Pathway

By adding at the above circuitry an inverter neuron which inhibits the output (Fig. **4.2**), one obtains a variant, the output of which fires only after simultaneous stimulation of both the inputs (activation) and only when the conditional input is firing: a new pathway has been opened.

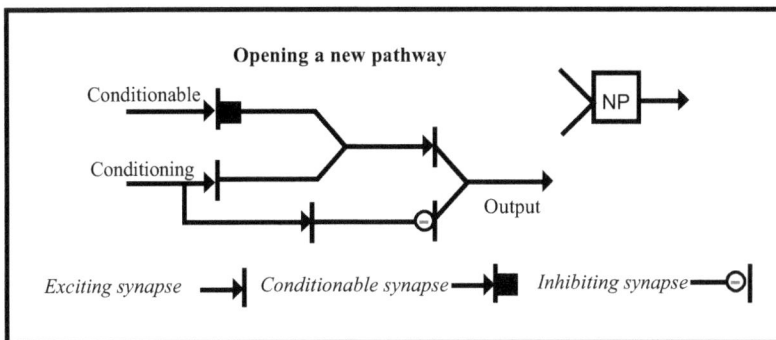

Fig. (4.2). Opening a new pathway.

After learning, the output fires if the conditional input is excited alone and only then.

Before learning:

    Conditioning neuron alone does not fire.

    Conditional neuron alone   does not fire

    Both neurons   does not fire, delayed n_R increase

After learning

    Conditioning neuron alone does not fire

    Conditional neuron alone   output neuron fires

    Both neurons   does not fire

From now, we will use the symbol NP (for New Pathway) to sum up the five-neurons wiring. Whenever the conditioning input is firing, it inhibits the output. Simple link and opening of a new path are stones used by memory and learning at the global level.

Why the Name "New pathway"? We will deal in 4.2 with many input neurons linked by NP junctions to many competitive output neurons. There are a huge number of possible pathways. Before learning, all of them are frozen. Once learning has activated one of them, the competitive inhibition acts to forbid later activation of the others: a new pathway is created.

4.1.1.4.3. Another Variant: Chronic Pain

Initially, a weak excitation of pain sensors is felt as a light pain at the CNS level. But, after a long and strong excitation (the learning period), a weak excitation is felt as a strong pain. This phenomenon is called chronic pain.

Pain sensors messages are transmitted to the spinal cord by aperiodic neurons and from this relay to the CNS (Appendix E). The normal output of the relay is an aperiodic neuron: the answer to a spike is a spike. Chronic pain appears when an activated glutamate synapse excites a periodic neuron. Thus, after learning, the answer to an isolated spike is a salvo. From this time, a short painful stimulus is felt as a long lasting pain.

4.1.2. Self Referent Regulations

Main facts: Slow regulations control the sleep duration or the end of thirst drive.

4.1.2.1. At the Molecular Level

The long term changes results from a very slow feedback. Pre-synaptic receptors in serotonin synapses or slow (metabotropic) glutamate receptors release messengers which reach the genetic apparatus of the pre-synaptic neuron and induce an increase of the synthesis yield of transmitter or receptor molecules [5]. The passive transport of the messenger molecules along the axon is a diffusion process with a coefficient about 1 cm^2/day. Its duration depends on the axon length. It is about one month for the below self referent end signals, about one day for focalized attention (5.1.4.3), about one minute for bees (4.2.1.1). Long term depression, the inverse phenomenon (slow feedback decreasing the synthesis yield of transmitter) is also known.

4.1.2.2. Self Referent Memory at the Behavioral Level

This phenomenon explains some rudimentary accustoming in primitive animals as aplysia [6]. In vertebrate, it plays a major part to adapt the logistics to the needs (by increasing the supply of molecules in frequently used neurons).

4.1.2.2.1. Self Referent End Signals

We have seen in 3.2 that drinking stops when a sensor located in the pharynx count a critical number of gulps. This critical number does not depend directly of the instantaneous need (the osmotic pressure of blood). We assume that it is specified by a self referent feedback: the critical number would be the mean number of gulps for the last weeks. Such an end signals govern probably the wanted quantity of food, the sleep duration and the sexual needs.

4.1.2.2.2. Self Referent Thresholds

The part of self referent feedback to settle the value of some thresholds explains perhaps why hard life increases the threshold of fear or pain feelings.

4.1.2.2.3. Obsessions

Obsessions (for instance the sexual one) are similar to long term impatience supported by long term excitation. Too frequent and too long instinctive drives (especially in the case of a lack of satisfactory end signal) induce an increase of the efficiency of the drive main neuron. Secondary drive outputs are recruited, which link with various behaviors: one observes a slow invasion of the most common signals by sexual connotations.

4.1.3. The Neonatal Plasticity (Epigenetic Events)

Main facts: Just after birth, some synapses need to be excited to become operational.

4.1.3.1. Connection of the Motoneurons to a Contractile Fibre

In the fetus, each fibre is reached by 4 or 5 motoneurons. After birth, neurons begin to fire. The most activated neuro-muscular synapse strengthens at the expense of the other synapses. Only the first one amongst other synapses becomes operational. Then, unconnected nervous fibres degenerate (an example of negative retroaction). Thus, sometime after birth, only one neuro-muscular connection remains.

4.1.3.1.2. Molecular Mechanism

This phenomenon was the first long term one to be understood at the molecular level [7]: short life non operational receptors of acetylcholine are synthesized during a short time after birth. The excitation of the synapse governs the transformation of the receptors into operational form. Attractive forces between receptors govern the kinetics of the phenomenon. The thermodynamic equilibrium leads to an asymmetric distribution with all the receptors captured by the same synapse (Appendix F).

4.1.3.2. The Visual Cortex of a Kitten

A similar phenomenon was observed in kittens. In adult men and cats, each pyramidal neuron of the cortical area 17 owns a narrow and precisely oriented visual field. In the kitten, each pyramidal neuron is excited by a distribution of visual field with a maximum in the final direction. The maturation obeys to the same mechanism as the neuro-muscular synapses. In normal conditions, competition eliminate all weak connections leading thus to a normal directivity of the visual field. But if the kitten is brought up in darkness, synapses do not capture receptor molecules and the kitten remains blind. If the kitten is brought up in a box with only vertical bright lines, only connections associated with vertical visual fields are activated and the kitten sees only vertical lines.

4.1.3.2.1. The Impression (Prägung)

It is a variant of the above phenomenon. Some birds have some hours after hatching to memorize the characteristics of their mother. Duckling records the first moving object answering to its calls and this object becomes its mother [8].

4.1.3.3. Remark

A slow degeneration of unused neurons persists perhaps lifelong. On the opposite, neurogenesis in adult brain seems mainly restricted to the hippocampus and the olfactory bulb. Unmasking secondary neuronal pathway is the most usual way to repair adult brain.

APPENDICES TO SECTION 4.1

A) The double gate of the NMDA channels

B) Glutamate link recording

C) Reading

D) Activation losses

E) The chronic pain

F) Molecular basis of plasticity

Appendix A: The Double Gate of the NMDA Channel

When released, the Glutamate molecule binds directly to an unusual channel owning a double gating: their opening Ω is the product of a chemical and of an electric term: $\Omega = \Omega_{CH} \, \Omega_{EL}$.

a)   The channel is selective, but has such a great diameter that it is crossed, when open, by K, Na and Ca cations.

b)   As for every swift synapse, the chemical opening Ω_{CH} of an activating receptor is 1 ms wide for each incident spike reaching the synapse.

c)   It is locked by external Mg ions when the local potential is negative. Now, this potential is positive only during the spikes. Ω_{EL} *is governed (as the more usual gates, see Section 1.2), by an equilibrium value Ω_{ELeq} depending on the potential V of the postsynaptic neuron and by a time constant τ. The equilibrium term is a sigmoid function significantly greater than 0 for V > 0. This happens only during a spike of the output neuron. The time constant value τ value is in the range 5 to 10 ms. Then, if the output neuron is firing with a rate around 200 s^{-1} (then, V is such that Ω_{EL} is continuously greater than 0) and if glutamate is simultaneously released (so that Ω_{CH} is large), then Ω takes a positive value for each conditional spike and Ca ions enter into the neuron.*

Appendix B: Glutamate Link Recording

When the output neuron is firing (then the electric gates of the special glutamate receptors are open), a fixed amount of Ca enters into the output neuron at each spike of the conditional one. *Each non operational receptor has then a probability p to become operational. There are 1- n_R non operational receptors. From experimental results, it takes 10 minutes to synthesize and transport the activating protein. Thus, 10 minutes after each spike, the average increase of n_R is:*

$$\delta n_R \equiv n_{(k+1)} - n_k = p.(1-n_R)$$

With the initial value $n_R = 0$, its value after k spikes is $n_{(k)} = 1 - (1-p)^k$. A reasonable value is p such that $n_{(k)} = 0.5$ after a one second, F = 200 s^{-1} salvo.

Then the values $n_{(k)}$ *versus* the number k of conditioning spikes is (Table **4.2**):

Table 4.2.   The proportion of activated synapse depends on the firing rate of the inputs.

k	20	60	100	200	400	600	800	1000
n(k)	0.07	0.19	0.30	0.50	0.76	0.88	0.94	0.97

Appendix C: Reading

The pavlovian link of Fig. (**4.1**) is supported by an aperiodic neuron, for instance the typical one (threshold $U_{Thr} = 6$ mV). Let F_1 and F_2 be the firing rates of the conditioning and the conditional inputs and, for the conditional pathway and let n_R (range $[0 - 1]$) the proportion of operational post-synaptic receptors. We assume weak synptic couplings: the effects of the two pathways are additive and the control potential is:

$$U = a.F_1 + b.n_R.F_2$$

where a and b are two constants. For instance: $a = b = 1/30$ (to obtain $U = 10$ mV when $F = F_{Max} = 300$ s^{-1}). The neuron fires if $U \geq U_{Thr}$

Conditioning pathway alone: $F_2 = 0$. Then $U = a.F_1$. The neuron fires if F_1 is great enough.

Recording: $n_R = 0$. Then $U = a.F_1$. The neurone fires if F_1 is great enough. If, at the same moment, F_2 is great enough, n_R will increase 10 minutes later.

Reading: $F_1 = 0$. The neuron fires if $n_R.F_2$ is great enough.

Appendix D: Activation Losses

We look at the spontaneous decay of the operational receptors population. *We have seen (Part I, Appendix 5) the general quantum formula for spontaneous decays:*

$Rate (A{\rightarrow}B) = (2\pi kT/h) \exp [(W_A - W_{Max})/kT$

Here, state 1 of receptors is their operational form, state 2 their non-operational form. W_{Max} is the height of the energy barrier and W_1 the energy of an operational receptor. We assume in this particular case the existence of repulsive forces between neighboring operational receptors. Then $W_A = w.n_R$ where w is the elementary repulsive energy. (We took the energy of an isolated operational receptor as energy zero). Then, with time in second:

2.26 10^{-12} dn$_R$/dt = n$_R$ exp(-W$_{Max}$/kT) exp(w.n$_R$/kT)

This equation gives the time as a simple integral of a function of n_R:

$$t = a\int_{n}^{1} \exp(-bx)\frac{dx}{x}$$

To determine the parameters a and b (W_{Max} and w), we own only two rough experimental results: a) Reading without any help of the conditioning neuron (that is $n_R > 0.7$) can be observed more than one day after the learning. b) Remembrance lasts lifelong. These conditions lead to a = 5.4 10^{10} (W_{Max} = 32 Kcal/mole), to w =13 **k**T (about 8 Kcal/mole) and to the values of Table **4.1**.

Appendix E: Chronic Pain

Here (Fig. **4.3**), the conditional neuron is periodic (while it was aperiodic in the above example). In the initial state, $n_R = 0$. Isolated spikes in the C fibre create in the slow synapse a messenger concentration and a conditioning control potential in the conditional relay $U_{Abs} \ll U_{Thr}$: the conditional relay remains silent and the response of the output neuron is an isolated spike.

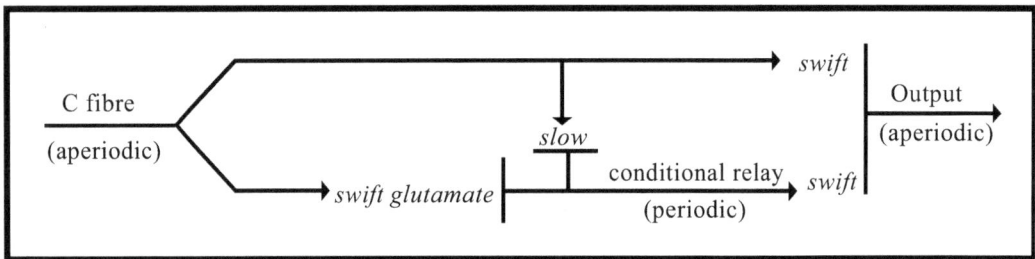

Fig. (4.3). A mechanism for chronic pain.

Long and strong (high firing rate) firing of the C fibre creates in the conditional relay an excitation $U_{Abs} > U_{Thr}$ while spikes reach the special glutamate receptor: some 10 minutes later, n_R increases.

Once the link is created, an isolated spike of the C fibre makes the conditional relay to fire. As it is a periodic neuron, it induces in the output neuron a periodic response (felt by the CNS as an intense pain).

Appendix F: Molecular Basis of Neo-Natal Plasticity

Neuro-muscular synapses use swift receptors of acetylcholine (called nicotinic). Free receptor molecules are produced at the ribosomal level and quickly decay (in about 1 day). The concentration c_F of free receptors is proportional to the production yield Y.

The yield Y is strong just after birth, and then it is very weak. Moreover, it is inhibited by Ca ions entering into the fibre when a synapse works.

Free receptors are not operational. Operational receptors are bound to the membrane in any of the 4 or 5 competitive neuro-muscular junctions. The populations of bound receptors in junctions 1, 2…will be called n_{R1}, n_{R2} …

There are exchanges of molecules between all the populations (Fig. **4.4**).

To study them, we look once more to the law of spontaneous decay of each population (which leads to the equilibrium conditions) with two particular assumptions:

a) *The energy barrier W_{Max} is lower when acetylcholine is present (it catalyzes the reaction).*

b) *There are attractive forces between neighboring operational receptors. Then, if zero is the energy of free receptors, the energy of bound receptors will be for each site $W = -w_0 - n_R \, w_A$.*

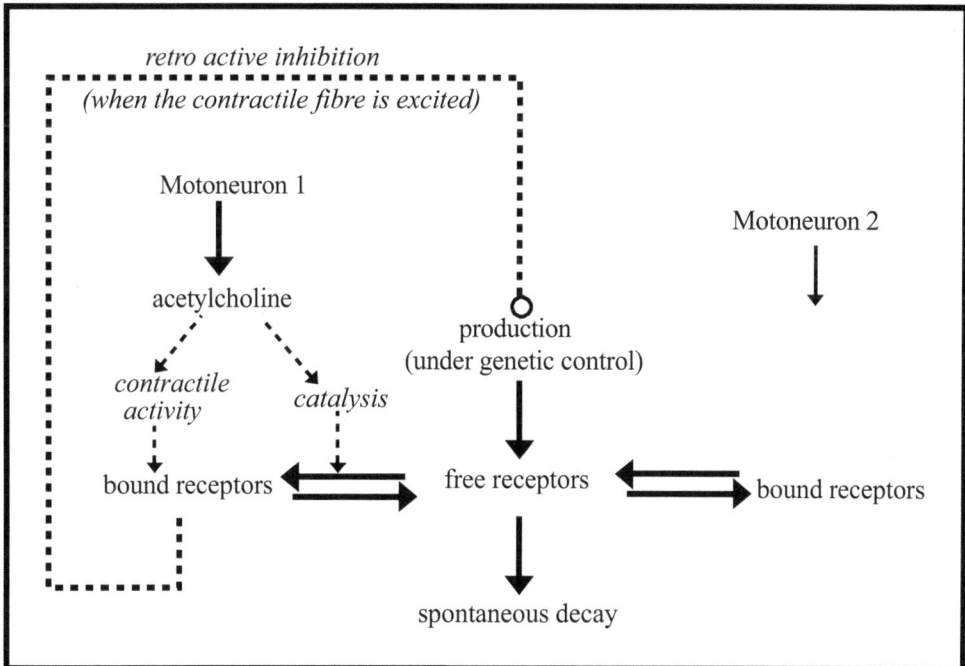

Fig. (4.4). Neo-natal destruction of unused synapses.

Working of the System

In the initial state n_R is small. If the yield Y is strong, concentration of free receptors c_F is high and the equilibrium values of n_R are not negligible. If the synapses are excited, the transmitter auto-catalyzes the exchanges and n_R increases till its equilibrium value. If the synapses are not excited n_r does not increase.

*Later, the yield Y and the concentration c_F decrease. If n_R had sufficiently increased before, **W** is so strong that great n_R remains in equilibrium with very weak c_F. But if the synapses have not be used (case of the kittens in darkness) the equilibrium value of n_R becomes awfully weak: the synapse will never become operational.*

Look now to the competition between two areas in the simplest case of weak yield. In the global balance, we can neglect the free receptors. Then we search the thermodynamic equilibrium with a constant total population: $\mathbf{n_{R1} + n_{R2} = n_{Tot}}$. The equilibrium is such that the sum of the binding energy $\mathbf{exp(n_{R1}/kT) + exp(n_{R2}/kT)}$ is maximum. The solution is either $n_{R1} = n_{Tot}$, $n_{R2} = 0$, or $n_{R2} = n_{Tot}$, $n_{R1} = 0$: any initial dissymmetry between synapses is amplified in such a way that all the receptors go towards only one synapse. And we have seen that unused synapses disappear.

4.2. Learning and Memory at the Systems Level

4.2.1. The Part of Modulations

Main results: Modulations govern recording and reading.

4.2.1.1. Spreaded Out Messages and Short-Term Memory

4.2.1.1.1. Present Time Duration

Various sensorial or emotive signals (for instance the pleasure impulse) combine before exciting a pavlovian link. In most cases, slow synapse and periodic behavior of neurons spread out the signals without memorizing them. A well known example is the retinal persistence value ($T_D = 0.03$ s), used to design cinema or TV apparatus.

The ***present time duration*** T_{PT} is related to the time constant T_A of slow synapses. Messages occurring during this delay can be added and globally interpreted. Simple psychological tests lead to the value $T_{PT} = 4$ s. Present time duration is a structure of our thought. Duration of verses (such as a couple of French alexandrines verses), melodies and explanations are related to T_{PT}. This limitation explains the empirical Shannon's law: in a stabilized language (English, French…), the length of words varies as the inverse of the logarithm of the frequency of their employment. (It is the code which maximizes the information transmitted during the present time duration). In the same way, it explains why mathematicians are obliged to introduce intermediate results to slice up the demonstrations while software accepts very long algorithms.

4.2.1.1.2. Short-Term Memory

Between the very short duration due to the synaptic delays and the infinite duration of long-term memory, the brain use the ON/OFF property of periodic neurons to generates salvos persisting some minutes or some hours. (One of the functions of sleep is perhaps to rub them out). Two modulations are used to start and stop the firing of the periodic neuron (Appendix A). Note that such devices have been experimentally observed. For instance, when a rhesus monkey is frustrated by delaying its meal (food hidden for some minutes), firing of a periodic memorizing neuron in the frontal cortex has been reported [9]: it begins to fire when a banana is shown and stops when the banana is found.

A variant of this system is a repeating device (Appendix B): in response to a starting impulse, the output fires every 20 s. (When drowsy, we repeat untiringly the same sentence).

4.2.1.2. Selective Reading by Long-Term Memory

We look now to the reading of a set of competitive pavlovian links. All the neurons are excited by the same slowly increasing conditional signal. Only the more excitable link (the more recently excited) will fire. But this ability to discriminate precedence is limited: that discrimination is very efficient for recent records, but not for old ones. Old remembrances reappear (Appendix C).

4.2.2. From the Honeybee to the Pavlov's Dog

Main results: The complexity of the simplest pavlovian phenomenon.

4.2.2.1. Memory of Flowers in the Supplier Honeybee

NMDA memorizing devices are supported by an inborn architecture, the connections of which are activated by learning. In the simple scheme of Section 4.1, only one possible input is linked to only one possible output. Such a device would be useless. In the honeybee, one among several possible inputs is associated with only one possible output. The simplest conditioned reflex links one among several possible inputs to one among several outputs. While the simple scheme uses only one NMDA synapse, the more sophisticated devices use many of them.

Gathering our knowledge of the phenomena at the cell level and of the observed behaviors, we are able to suggest a plausible circuitry. Its ingredients are short-term memory, activation of simple NMDA link, opening of a new pathway, use of choosing boxes (several competitive inputs, only one output firing at a time).

4.2.2.1.1. The Honeybee Behavior

A bee owns only 10^7 neurons. In summer, a worker lives about one month long. During this time, it is successively a nurse, then a builder, then a supplier. Von Frisch, circa 1910, described the behavior of a supplier. Just after its transformation (induced by some hormones) from builder into a supplier, the bee stays in the beehive and shows signs of a strong excitation. A former supplier comes back and dances. The young one records the hour of the dance, the angle between the axis of the dance and the axis of the beehive, (which codes for the angle between the direction of the flowers and the direction of the sun) and the amplitude of the dance loops (which codes for the distance of the flowers). The bee flies to the indicated spot, find a first flower. Then, it records the color, the shape and the smell of this specific flower [10]; it gathers some nectar and comes back to the beehive, first by using the sun direction, then by using of the smell of the hive as of a beacon. The following days, the bee flies at the same hour towards the same flowers. When the seam is exhausted, the honeybee dies.

4.2.2.1.2. The Circuitry for Memorizing Flowers

First, we look first only to the color learning. Inborn sub-units process the visual input to excite color sensitive neurons, which act as competitive conditional inputs of the learning device. Several color inputs C_1 (purple), C_2 (yellow)… are the conditional inputs of new pathway devices NP_1, NP_2… (New pathway devices were described in 4.1). The number of possible competitive inputs is small (under

10): look at bee gathering nectar of multicolored lupines: the supplier despises red, pink or yellow flowers, but it gathers mauve and violet lupines: it does not distinguish these too slightly different colors. A first branch of the emotional excitation due to the dance acts as a conditioning modulating signal. Simultaneous excitation of a conditional input (a red flower) and of the conditioning one activates a link with one of the competitive new pathways. The synaptic couplings between any NP box and the common output is strong enough to make the common output firing (operation OR). A similar scheme with AND operations will be described in 4.2.3.

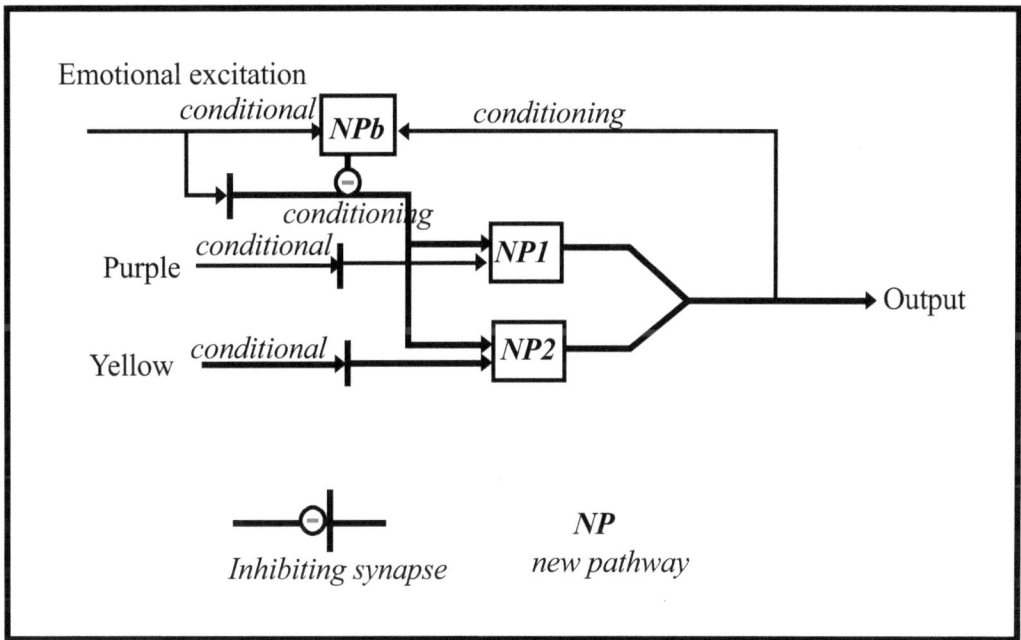

Fig. (4.5). Memory of flower color in a honeybee.

The circuitry has to be completed to avoid the later recording of a second color. The new output and the modulating signal are used to open another new pathway, which exerts an inhibiting retroaction on the modulation (Fig. **4.5**). The first color seen while the emotional excitation is firing is linked to the output. (In realty, the flower signature is its color AND its shape AND its smell). From this moment, the inhibiting loop is activated, the emotional excitation cannot reach its target, and other colors seen later are not linked to the output. Such **bolting** retroactive new pathways will be found in conditioned reflexes.

4.2.2.2. The Pavlov's Dog

4.2.2.2.1. The Experimental Facts

The experiment has to be done at meal time, when the dog is hungry. The dog hears a bell a short time (about 1 minute) before receiving food. A conditional reflex appears: after some repetitions, the bell, at meal time or at another time, triggers gastric juice secretion.

The main phenomenon obviously is the activation of a conditional link. But, as demonstrated by electro-physiology, the macroscopic conditional reflex is a much more elaborate phenomenon that the change of the properties of only one neuron; several links seem to be created in various places of the brain. This is due to the difficulty to associate non simultaneous signals and on the other part to the complexity of commutation (several possible inputs and several possible outputs have to be linked).

4.2.2.2.2. The Actors

a)   Hunger is one of the main competitive instincts. At meal time, the hunger drive increases, this instinct is chosen. Pre-eating behavior (salivation, gastric juice secretion) is triggered. The end signal induced by the food (the reward) stops the hunger drive and creates a pleasure impulse.

b)   When the bell is ringing the first sensorial areas generate a set of signals (associated with the duration, the intensity, the location of the sound, its main frequency and the relative intensities of its harmonics).

The simplest conditioned reflex (this Section) links one of these signals to the hunger drive. Differentiation (next Section) links several of them to the drive.

4.2.2.2.3. A Simplistic Scheme

A switching device (for instance the web) is needed to link several phones, several computers or several neuronal lines. As a first step, we look at a simplified scheme (Fig. **4.6**). Among all the sensorial signals S_1, S_2 …, only one (Σ_1) is chosen after going through a competitive choosing box. The bell is heard some minutes before the gift of food: each output of the choosing box excites a short-term memory supported by a periodic neuron P. The system needs an exciting modulation; the most obvious is the hunger drive. When food is obtained, the hunger drive end signal generates a pleasure impulse which acts as a conditioning signal (often called the absolute signal) while the output of the short-term memory acts as a conditional signal to open a new pathway (NP) toward the hunger drive circuitry. From this moment, hearing the bell is sufficient to initiate salivation. (The end signal stops also the periodic neuron after a short delay).

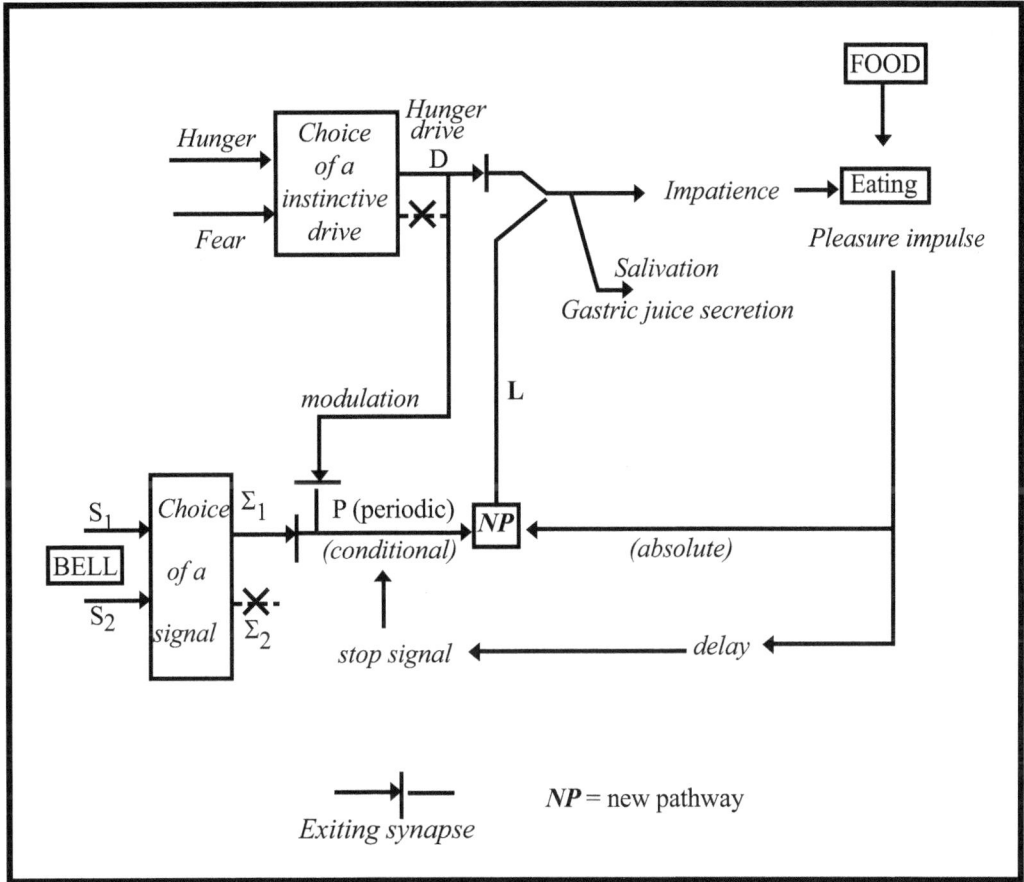

Fig. (4.6). Simplistic scheme of Pavlov' reflex in a dog.

4.2.2.2.4. A More Realistic Scheme

We face now a commutation problem similar to the one well known by engineers: conditional reflexes can link almost any elaborate signal to almost any elaborate instinct (for instance, the view of a flashing red spot can be linked to the pushing of a footbrake). The number of such possible combinations is almost infinite. To economize the number of neurons, the system uses an intermediate sub-unit of limited size, which can support a limited number among an infinite number of possible combinations. (We encountered such a device when studying the sense of smell. Its benefit of it will be discussed in the next Section).

The above simplistic scheme has to be slightly modified (Fig. **4.7**; changes are pointed out by gray zones) to include 4 simple links or NP devices for each reflex.

a)   The exciting modulation of the short term memories is not the hunger drive, but a general emotional signal (obtained by some OR operations) associated with any high frequency instinctive drive (associated with hunger OR fear OR sexual desire OR…). In the same way, the conditioning signal is not the pleasure impulse associated with eating, but a signal associated with any strong reward.

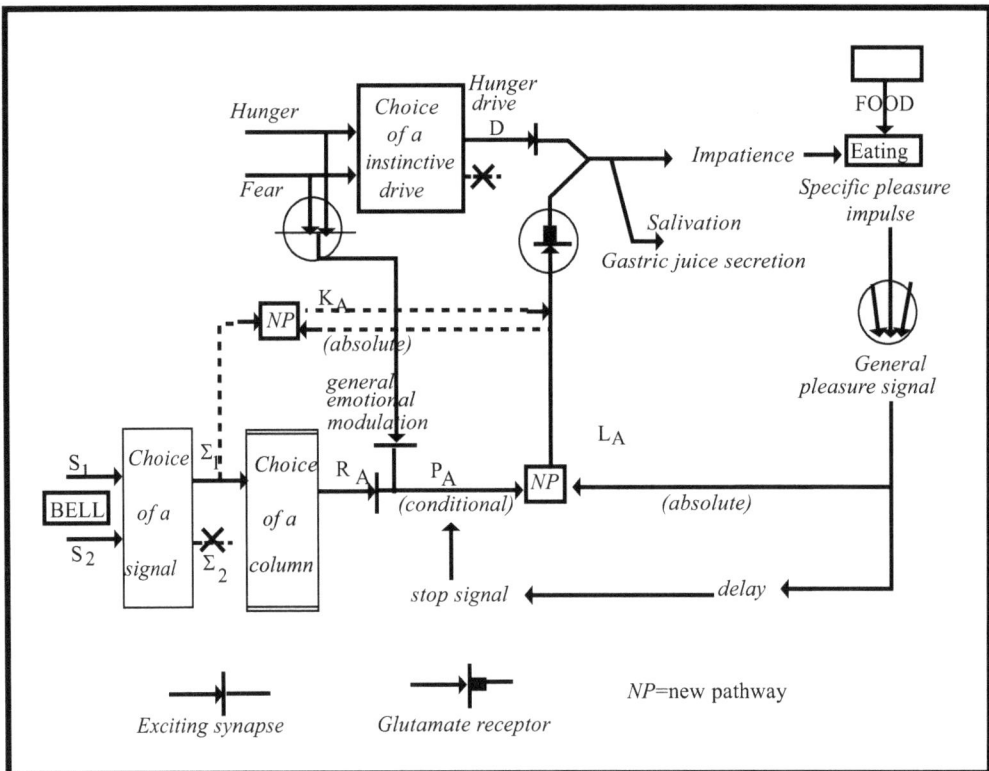

Fig. (4.7). A more realistic circuitry.

b)   The periodic neurons of the short term memories are not connected in an inborn way to each of the output of the choosing box. The selected signal Σ_1

(for instance the duration of the bell sound) reaches a second box, which chooses a column (index A) made of an output R_A, a periodic neuron P_A, a new pathway line L_A. This new pathway is not associated in an inborn way to the hunger drive. The simultaneous excitation of L_A (acting as a conditional input) and of the hunger drive (acting as a conditioning signal) activates a new link with the drive.

c)   To preserve the specificity of the association between an input Σ_1 and a column L_A when the device is submitted to other different tests, the links have to be bolted (schemes in Appendix D). After bolting: associations of Σ_1 with another column L_B or of L_A with another signal Σ_2 are forbidden (but the activation of a link $\Sigma_2 L_B$ is allowed). On the other hand, a direct and fast pathway (dotted lines on Fig. **4.7**) goes from Σ_1 to L_A bypassing the second choosing box (slow motor responses of learning drivers are replaced by very fast responses after learning).

4.2.2.2.5. Remark on Rewards

In Pavlov's dog, the contact with the food triggers the end of drive signal and the pleasure impulse. More generally, any signal triggering end of drive and pleasure impulse plays the part of a reward. It can be a material reward (the food), a social signal (when the master pats its dog) or a purely mental signal: remembering the name we were searching ends the recalling effort and is felt as a pleasure.

4.2.2.2.6. Remark on Emotional Loops

In many cases, a primary sensorial signal (or a more elaborate one) link with a behavior, which acts on the deeper layer of the brain to induce an excitation of the attention. The attention excites the thalamic area, which modulates the cortical area analyzing the signal. Such a learned loop increase the sensitivity to this peculiar signal: we spot a beloved face in the middle of a crowd. In some other cases, the learned loop is inhibiting: we avoid annoying subjects. The extreme case is freudian taboo.

4.2.2.2.7. Remark on Cortical Re-Mapping

It was observed that the brain activity associated with a given function can move to a different location as a consequence of normal experience or brain damage/recovery [11]. The spontaneous decrease of $\mathbf{n_R}$ of the inhibiting links in

Fig. (**4.7**) when the device is not used for a long time allows the recording of new signals.

4.2.2.2.8. The Part of the Signal Intensities

A child sets fire to the tail of a cat. After this joke, the animal will never let the child to approach it. The remarkable fact is that only one test was sufficient to noxious stimuli to build up a reflex while other reflexes need several repetitions. We show in Appendix E how this quickness is due to the persistence and the strength of the emotional behavior (the fear). On the contrary, we note in the same appendix that weak rewards induce very slow conditioning processes. This property is needed to understand differentiation.

4.2.3. Differentiation and the Building Up of Concepts

Main result: The building up of abstract messages.

4.2.3.1. The Mechanism of Differentiation

Differentiation is a progressive refining of sensorial analysis: for a child, all Chinese are alike; some years later, he is able to differentiate between them. Differentiation is obtained by adding AND logical operations to the Pavlov's mechanism. A variant of differentiation is the activation of sequential links. Abstract concepts are obtained by adding operations OR to the above mechanism. Several abstract concepts are often associated one on the top of the other, leading to very subtle identifications (for instance when an open or a closed red or black object is recognized as an umbrella).

4.2.3.1.1. The Experimental Facts

A main result of Pavlov's experiments is the discovery of differentiation [12], a phenomenon enabling progressive learning. Once a conditional reflex has been created, its working is tested with the original bell and with another bell of slightly different sound. You observe the same secretion of gastric juice for both the stimuli. From this time and in a systematic way, food is given when the first bell has been heard and never given when the second bell has been heard. After some days, gastric juice is secreted after hearing the first bell, not after hearing the other: the dog has learned to discriminate its stimuli.

4.2.3.1.2. A New Actor: The Resentment Impulse

Differentiation is obviously due to disappointment (no food received). We have seen in 3.2 that disappointment generates a strong modulating pulse, symmetric of the pleasure pulse. We called it resentment.

4.2.3.1.3. A Multi-Floors Device

The differentiating device is made of several floors (6 for sense of smell, see 3.1.1). In this case, the same set of sensorial signals S_1, S_2... excites all the floors. Each of them is almost similar to the simple reflex device (Fig. **4.7**) with two choosing boxes and an intermediate sub-unit and only two differences: a) through P_A and O_A the direct pathway S_1K_A inhibits L_A (Fig. **4.7**) and the transmission of S_1 toward the upper floors; b) the general emotional signal reaches only the first floor. In the second floor, its part is played by another general signal Λ (Fig. **4.8**).

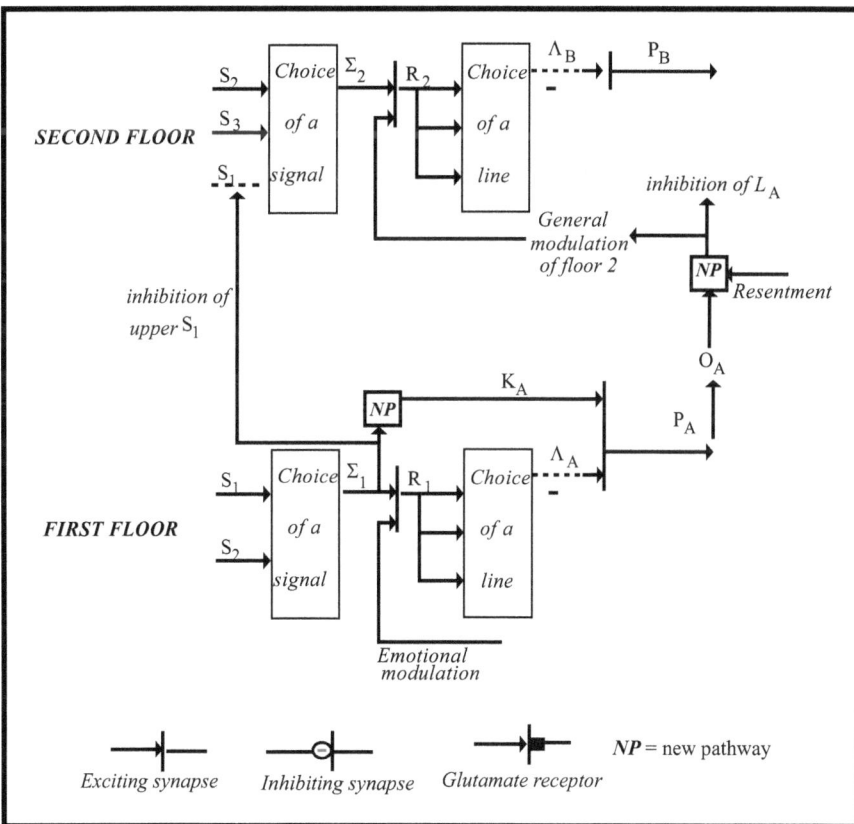

Fig. (4.8). A multi-floors circuitry.

The resentment impulse acting as a conditioning signal and L_A acting as a conditional one open a new pathway which a) inhibits the transmission of L_A toward the hunger drive and b) excites Λ. The second floor acts as the first one: bell + food induce the linking of a line L of the second floor with the hunger drive. Thus, sensorial signals S play the part of letters and the activation of 1, 2, 3 floors is equivalent to the building of passwords made of 1, 2, 3 letters. After learning, the reading of a correct word (the hearing of the good bell, the recognition of an odour) induces the firing of the upper floor output, which is a signal specific of the recognized word. We call it the label of this word. The label triggers the instinctive behavior (the gastric juice secretion).

4.2.3.1.4. How to Choose a Differentiated Signature; Part of Weak Resentments

In smell sense devices, the S inputs signals are ranked before reaching the cortex following their intensity. Then, the first floor recognizes the more intensive signal associated to some smell, the second floor the second intensive signal and so on: the signature of a smell is the list of its six more intensive signals. In the two bells experiment, things are less simple: is the signature bound to the first harmonic relative intensity or to the signal duration or to some visual complementary information? Weak resentments, because they produce slow learning, enable the building of convenient signatures.

Bell Without Food Delivery: Once the reflex has been created, the bell heard by a well-fed dog is never more followed by food. As the dog is well fed, resentment is weak; the activation of the device inhibiting the conditional link is very slow. As a result, the initial conditional reflex can be tested without being destroyed. After a lot of bell without food repetitions, the link is, at last, inhibited. The second floor is allowed to work, but has not yet found a target: at this time, hearing of the bell does not induce any observable response.

Searching for Correlations: In everyday life, the subject perceives a mixture of significant and of meaningless signals. How to find the significant ones? For instance, a well-fed dog receives food if a first bell is heard; it does not receive any reward when a second bell is heard. At the same time, the food provider wears randomly blue or red clothes. But this difference is not notified to the animal, it has to spontaneously learn to separate significant from noisy information. Progressive activation of the second floor prepares slowly the use of first bell and the use of red shirt as differentiated signals. But the systematic combination (first bell + food) occurs twice more often that the random combination (red shirt + food). So, the bell line is the first to be activated and prevents (inhibits) the activation of a red shirt line.

4.2.3.2. Sequential Linking

In the smell sense device, signals are ranked by their intensity: the highest frequency signal is chosen by the first floor box, the next one by the second floor box and so on. We know (Appendix J to Section 1.3) how to detect intervals between sequential events. A chain of such delayed lines memorizes a melody. In this case signals are ranked by the time of their occurrence: the musical motif "ta ta ta poom" will be recorded as "ta THEN ta THEN ta THEN poom". (In the circuitry, periodic neurons which fire continuously after being triggered are replaced by sequential detectors in the figure of this appendix, which fire after some fixed delay). Sequential links play a main part in the analysis of hearing and of motor control.

4.2.3.2.1. The Fate of the Bad Bell Signals

In the two bells experiment, hearing the bad bell induces the firing of some L neuron of the second floor (call it L_{BAD_BELL}) while neither hunger nor fear drives are firing. But attention induces some impatience, the end of it acts as a weak satisfactory signal. Then, the bad bell signature links with end of attention: it means bad bell recognition. More generally, the L_{BAD_BELL} signal links with a drive different of the original one. Thus, a recognizing tree is gradually built up. For instance, a baby links human face with food. After a first differentiation, mother remains linked with food while various social rewards lead to differentiate father from others, then brothers from others and so on.

4.2.3.3. The Building of Abstract Concepts

4.2.3.3.1. Multi Sensorial Associations

Learning areas build up associations of different types of signals (for instance, shape and color) or of multi-sensorial signals (hearing, smell and body position). While the first steps of the signals analysis are supported by easily located areas of the brain, multi sensorial associations are mainly supported by the frontal cortex (a part of the brain especially developed in man).

An Example: The Dog and its Owner: Wolves and dogs own inborn social instincts. They need a leader. In domestic animals, a man is chosen as the leader.

In a first step, the owners' smell is linked in a conditional way to the obedience instinctive behavior. One of the outputs of the behavior is an emotional attention, which will be used to create associations. In a second step, one observes the building up of associations between the owner's smell, his/her voice, his/her clothes and so on, each of them playing after learning the part of a partial criterion for recognition. In the same way, linking between gift of a piece of sugar, owner's voice intonation and pats on the head builds up new rewards.

Any Logical Operation Can Be Carried Out by a Sequence of Learned Devices:
Boole's theorem specifies that combinations of the three basic operations AND, OR and INVERSE can carry out any logical operation. We have seen that an inhibiting relay carries out the logical operation INVERSE. The output of a two floor differentiating device fires IF the first input AND the second one fire simultaneously. So, we are logically led to look at learned OR operations.

4.2.3.3.2. AND/OR devices

A simple variant of the conditional reflex circuitry (Fig. **4.9**) seems conveniently describe AND/OR devices. The output of a differentiating box (for instance, the owner's voice) acts as a conditioning signal. Simultaneous signals coming from other analyzing devices (the owner's smell, the color of his/her dresses) combine with conditioning one to open new pathways J_A, J_B... The main point is that all the J are connected to the common output O through weak synaptic couplings.

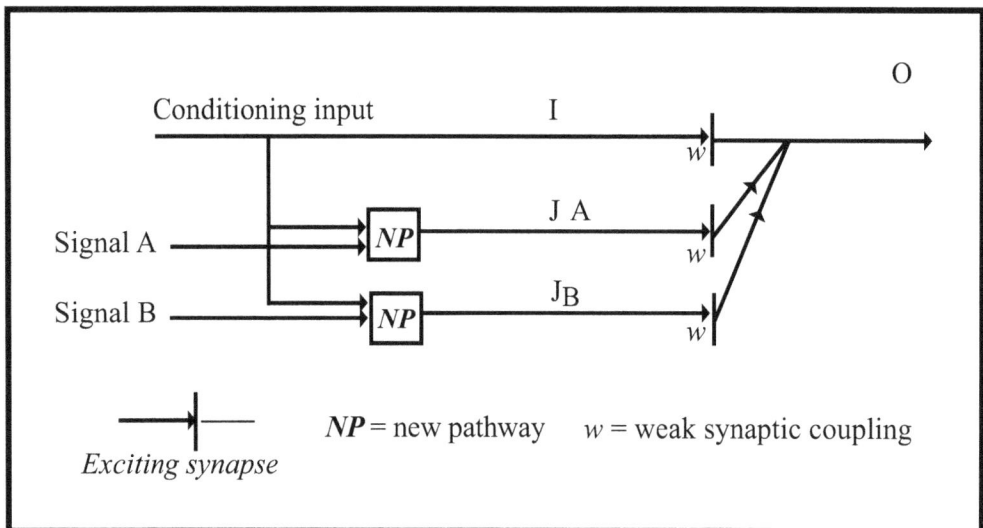

Fig. (4.9). An abstraction learning circuitry.

Properties: Before conditioning, all the J neurons are mute. The input I is too weakly coupled to the output O to make it firing. Simultaneous excitation of I and S_A causes the activation of the new pathway J_A. After learning, the excitation of S_A alone makes J_A firing. But the weak synaptic coupling makes O mute except if the sum of the control potentials of the firing neurons overtakes the threshold of the output neuron, for instance if (I AND J_A) OR (I AND J_B) OR (J_A AND J_B) are firing.

As a result, the device recognizes the owner even if he/her changed his/her perfume or the color of his sweater. (This is similar to the automatic recognition of a target by a device using fuzzy logic). Such conceptual output can be used as input into another associative device. The combination of successive associations and differentiations in several superposed learning devices have outputs firing as a response to the sight of a red or black, open or closed umbrella or as the signal "Me personally", which will be met in 4.2.5.2. Such an output label is characteristic of an abstract concept (which can often be designated by a word). It can be used to trigger a divergent process, for instance the detailed recall of old sensorial signals.

Note that this device is asymmetric: the conditioning input I (here, a hearing signal) attracts the others. We will see that, completed by backward neurons, associative devices play a main part in the cognitive functions.

Some associative devices have been located, for instance a device located in the left infero-temporal cortex, which associates the appearances of the same word written in small or capital letters or different characters [13].

4.2.4. Recollection and Imagination

Main results: Memorizing devices generate virtual (imaginary) sensorial signals

4.2.4.1. The Memory of a Nursery Rhyme

In its most usual meaning, the word "memory" applies to a device able to recall old sensorial signals. This device is a black box with sensorial inputs, real sensorial outputs and virtual sensorial outputs for each note, rhyme label outputs and evoking rhyme inputs specific of each learned melody (Appendix F).

4.2.4.1.1. Sensorial Inputs/Outputs

The sensorial inputs are not brute signals, but processed ones: these signals I_{AB} are linked to: "any transposition of (note A THEN note B)". Each input excites directly a sensorial output O_{AB}, which transmits farther the signal, for instance to trigger singing. (The tempo and the accentuation seem to be analyzed by a parallel device).

4.2.4.1.2. Structure and Function of the Memorizing Device

The device works on the side of the direct input/output pathway. It is in charge of three functions: to record, to recognize and to evoke. Recording and recognizing a melody are very similar to recording and recognizing a smell. Here, successive signals are analyzed by the successive floors of the device. Fig. (**4.10**) describes the simplified scheme of a floor.

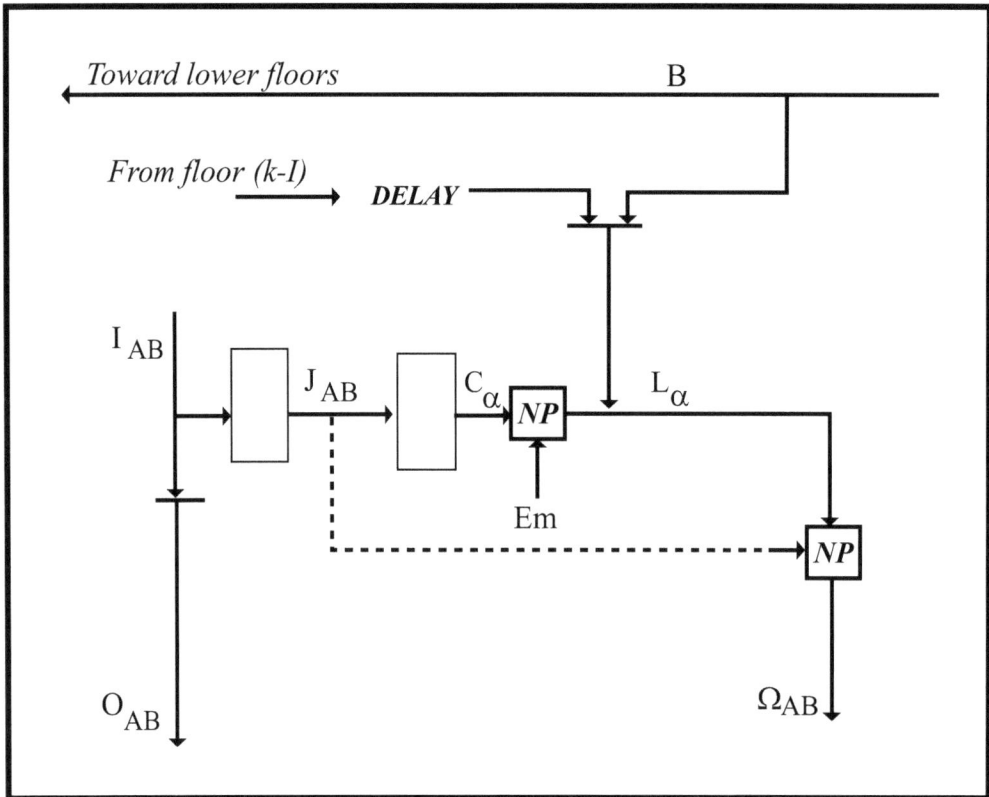

Fig. (4.10). Memorizing a melody.

The signal (I_{AB}) coming out of the primary sensitive area is a harmonic combinations of two successive sounds. The first choosing box detects (I_{AB}). The second choosing box associates I_{AB} with a still available line C_α. A modulating signal coming from the floor k-1 allows recording. The signal $L\alpha$ of the upper floor (similar to the address of a computer memory) is the label of the melody. It is associated with a musical theme and only one. A bolting device (Appendix G) preserves the specificity of the association of a label with a musical theme.

4.2.4.1.3. Backward Neuron and Virtual Sensorial Output

The main difference between a memorizing device and a differentiating one is the ability of the memory to recall old sensorial signals. This property is due to a backward circuitry built around a backward neuron B, going from the upper floor toward the lower ones and specific of the melody (as the label). When excited, it first induces the firing of the lower floor. Then, the B signal and a delayed signal coming from the floor k-1 induce the signal of the floor k (Appendix G).

When the neuron $L\alpha$ of the floor k fires while the real sensorial input I_{AB} is mute, the virtual sensorial output Ω_{AB} fires (dotted line in Fig. **4.10**). In many case, further processing separates the virtual character from the purely musical signal, which joins the real output pathway: a virtual duplicate of the original melody is generated. Note that a modulating excitation is required to fire the backward neuron: the effort to remember followed by a spreading of the arousal is called "the priming" of the evocation.

4.2.4.1.4. Size of the Memorizing Device

The number of possible *A* THEN *B* input signals can be evaluated: we distinguish 12 ratios in each octave. Looking at 5 octaves up and 5 down, the number of possible invariant primary signals would be about 120. (Is the Chinese music characterized by the same number?)

The present time duration gives an idea of the number of floors: less than 10 (for instance 6 as in the sense of smell device). Experiment seems to show that we are able to remember from 100 to 1000 songs. Then, the memorizing device would be made of 6 floors, each of them with a number of columns $10^2 < n < 10^3$.

4.2.4.2. Evoking and Imagining

Several schemes combining label generating devices with backward neurons B and associative AND/OR devices enable the building of an amazing variety of abstract virtual signals. (In most of cases, transmission of some virtual signals toward the corresponding real sensorial outputs is inhibited. Impaired inhibition leads to hallucinations).

4.2.4.2.1. Spontaneous Recollection

We have seen (Appendix C) that a slowly increasing impatience exciting all the competitive inputs of a learning device triggers the reading of the last excited pathway. The same process applies to the set of N backward neurons of a sequential memory: one pathway is triggered and the virtual sensorial outputs of the corresponding rhyme fire one after the other. Thus, a rhyme can be recollected without any sensorial triggering.

4.2.4.2.2. Backward Neurons Inside Associative Devices

Backward neurons are added to the associative device of Fig. (**4.9**) in charge, for instance, of a musical theme OR rhythm association (Appendix H). The musical theme alone excites the common output. A backward neuron triggers then the recollection of the associated rhythm. (A difference between this recolled rhythm and the really heard one causes some trouble. This is the principle of some musical parodies).

4.2.4.2.3. Pyramidal Recording Devices

To explain the memorization of long-duration messages, we must assume a pyramidal hierarchy of memorizing devices. The labels of short musical themes can be used as the sensorial inputs of the next hierarchical level. Its output is the label of the couplet. The outputs of successive couplets are linked in a higher-level device. Its output is the global label of a song.

Remind that the duration of the NMDA activation cycle is about 10 minutes. This delay explains that the best way to learn a long text or a complex maneuver) is to acquire separate fragments of the pyramid one after the other by short learning periods separated by more than 10 minutes.

To Summarize: Combinations of memorizing and associative devices allow the almost spontaneous generation of long sets of virtual, imaginative signals. We will see that short term imagination plays a part in motion control and long-term imagination in human thinking.

4.2.4.3. The Marcel Proust' Madeleine

Marcel Proust wrote a famous description of recollection: some afternoon, the old writer dips by chance a special cake (a "madeleine") in his cup of tea. He feels a strange trouble. After a very long and painful effort, he remembers he dipped such cakes in tea when, twenty years before, he visited his aunt Leonie in Combray, the small town where she lived. And then a whole lot of memories of Combray come back to him.

This experiment is obviously a case of associative memory. The taste of a madeleine in a cup of tea was such a rare sensation that its relation with Tante Leonie had not been masked or inhibited by rivals. But it was weakened by time. From Table **4.1**, we learn that, after twenty years, the number of activated channels is only 20% of their original number. So, restoring the connection required extremely strong modulation. And, because the deactivated channels are not reactivated when the neuron is excited but some ten minutes later, the modulation has to act during more than ten minutes. Such an excessive effort to remember had to be voluntarily maintained and was resented as painful.

When the first memorizing device had been reactivated, its label was used to excite upward the pyramidal structure (linking the taste of the cake to the abstract concept of Tante Leonie), then backward to link the general abstract concept to a lot of detailed virtual sensorial outputs.

4.2.4.4. Middle Term (Working) Memory

4.2.4.4.1. Experimental Facts

We remember easily what we have done yesterday, but not what we have done a month ago. So, we have at our disposal a middle term memorizing device that works when short and long term devices are not activated. (Note that the three types of memories are similar to the register memory, the RAM and the disk of a computer). The re-writable middle term memory is usually called a "working memory".

Working memory devices have been localized in the dorso lateral sector of the prefrontal cortex of primates, men and rats [14, 15].

4.2.4.4.2. Inferences

The usefulness of a middle-term memorizing device is clear: the number of stored data in any device is equal to the number of data daily stocked multiplied by the storage duration. So, an available number of neurons enable either the stocking of many everyday data during a limited time or the storage of few everyday data during a long time. Detailed data are kept some days in the working and only a few of them are transmitted to the long term memory.

Periodic neurons are unable to fire for some days. Thus, working memory has to use glutamate synapses in some variant of the circuitries described in the above Sections. We have seen that the conditional reflex circuitry owns a bolting device allowing the exclusive attribution of a line to a sensorial signal (Fig. **4.3**). Without bolting, new links rubs out the older. So, we assume that working memory is supported by circuitries without bolting.

4.2.4.4.3. Medium-Term Memory and Ageing

If the neuronal metabolism is slowed down, for instance owing to vascular troubles, activating new NMDA receptors (a highly fuel consuming process) becomes difficult, but already activated receptors are not deactivated: the first symptom of this kind of troubles is the disappearance of medium-term memory.

4.2.5. Learning, Conceptualizing and Executing Motions

Main results: Motor imagination. "Me personally" signal. Imitative learning.

4.2.5.1. The Exploring Instinct

Most of our usual motions (for instance, driving my car) result from a long learning process. Animals (and above all young animals) do not wait passively for an occasion of learning. They own a specialized inborn instinct, the function of which is to provoke the occasion. Its drive is a chronometric one, which is a self generated need for bustling daily during some fixed time.

Without any external stimulus, primates investigate the strange gadgets put in their cage; random walk is used by rats to explore a labyrinth without any apparent benefit. In man, the exploring instinct is not only curiosity, but need to act and feel. Complete silence is painful; a spontaneous drive increases until we obtain a convenient level of background noise. In the same way, we need tactile feelings: this is the usefulness of rosaries.

Exploring instinct is the first motor of learning.

4.2.5.2. Babies' Individual Learning

When babies, more than 2 months old, explore their motor functions their exploring drive induces random muscular motions without any external stimulus. For instance, the exploration of the phonetic motor apparatus (the babbling) is a random exploration of the phonetic motor apparatus. Self hearing generates a pleasure signal (end of the drive and reward). It leads to the linking between motor order and the sensorial signal occurring after the order. Differentiation then separates sensorial signals following an order from sensorial signals without order: the baby distinguishes himself from the outside world. Associations OR between the various self created noises converge toward a common output, the firing of which has the meaning "**Me personally**". On the other hand, virtual motor orders associated with the real ones link with the expected sensorial signals, building **imaginary responses to virtual orders.** Very often, the imaginary information is divided between information without indication of its realty and a real or imaginative yes or no signal.

One result of the motor exploration is a purely tactile and mechanical representation of the body (see 3.2.2.2).

4.2.5.3. Babies' Social Learning

All animals recognize animals of the same species. The universality of children drawings shows that human race is recognized by the association of a head, a trunk, two arms and two legs. Recognition goes in each species with an inborn social behavior supported by a communication system. The simplest one uses pheromones. Dogs use a triple system: a) they urinate to mark their territory and the smell is perceived by the other dogs; b) their emotions govern the position of their ears, their lips and their tail; this code is recognized by the other dogs; c) their barking express their main emotions; the barking is heard and understood by the other dogs. The inborn communication instincts of man are recognizable by their universal character. Facial expression, some stereotyped gestures and shouts express our main emotions.

Babies (8 to 11 months after their birth) begin to use this proto-language to establish associative and differentiating links between themselves and other people. From a) the mechanical representation of our body and b) the visual characteristic (as shown by children drawings) of another man, we build up a visual representation of our body. The outputs of this device can be combined with the corresponding outputs of another man representation and the difference can be used as a (real or virtual) motor order: this is the **basic mechanism of imitation.**

4.2.5.3.1. Men and Dogs

The communicative signals of a dog (gestures, facial expression, modulated barking) is easily understood by its owner. On the other hand, some signals of the owner (voice intonations, some words) are learned by the dog. Both the sets of signals make up a dissymmetric communication system very similar to a proto-language. They enable social dog-man relationships.

4.2.5.4. The Motion Controlling Devices

There are some inborn stereotyped motions. But most of usual motions result from a modification of inborn mechanisms or are fully new. With time, many of them become learned stereotyped motions (for instance riding a bicycle or lifting a spoon to our mouth), carried out quickly and without hesitation. In other cases (such as jumping from a rock to another), the real motion must wait during a sequence of virtual trials.

Each particular global motion is governed by a specific order acting as an instinctive drive. The competitive global orders of all possible motions compete in a choosing box. Only one global motion is chosen. Global motion orders act as the label inputs of a sequential memory device. The backward neurons build up a sequence of detailed orders, which are directed toward real or toward virtual muscular pre-orders. This part of the process is supported by the supplementary motor cortex. Virtual orders are processed in the pre-motor cortex: a trial and error correcting device enables, when needed, to test the coherence and the effect of a motion before releasing it (for instance, it adjusts the muscular strengths before jumping). In some cases, this device is bypassed by a fast direct pathway. At last, detailed orders are dispatched in the motor cortex and transmitted to the various layers of the motion governing devices.

We look at the example of car driving once learning has been carried out. The motion is triggered by an instinctive drive. One branch of this drive excites the choosing box. The output of the box is the global order. Other motions are forbidden (when driving, walking is inhibited). The global order creates a global modulation, which stirs up the specific wheel control. Another branch of the instinctive drive is associated with the choice of a wanted direction. In addition, it imposes a sitting posture, which is detected by body sensors to create an "I am driving" signal.

The turning wheels control works when and only when it is excited by the global modulation. Wheels control is embedded in the following loop (Fig. **4.11**): Landscape → visual signal, which is compared with a wanted signal; the difference → muscular preorder → effective muscular order to turn the wheel → reaction of the car → changes of direction → changes in the landscape. The figure contains an imaginative loop starting from the virtual outputs of the supplementary motor cortex. For the present paragraph, we assume that the feedback gain (how much to turn the wheel) has been fitted to stabilize the car and the imaginative loop is bypassed.

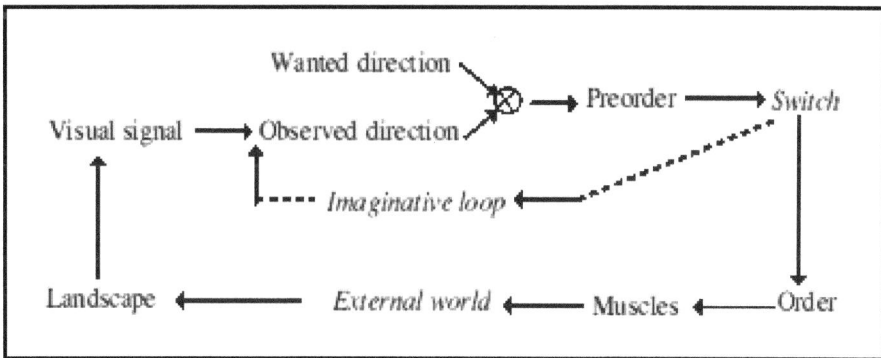

Fig. (4.11). Turning the wheel of a car.

More generally: When the global modulation fires and the imaginative loop is bypassed, real muscular signals are used by the external loop. When the imaginative loop is not bypassed, real muscular orders are inhibited, an imaginative visual direction change is predicted from the turning wheel preorder. If compared with the wanted signal (to stay on the road), it enables imaginatively test and to plan complex maneuvers. When the imaginative loop reach a satisfactory value, the switch position changes and real muscular orders are transmitted.

4.2.5.5. Some Examples of Motion Learning

4.2.5.5.1. Training a Wolfhound to Attack

Fighting is an inborn instinct of the dog. Some characteristics of the word "attack!" (its intonation, its length…) are used as conditional signals to activate a simple reflex. The reward is either the defeat of the foe or the satisfaction of the master.

4.2.5.5.2. Training a Poodle to Set Up and Beg

A child trains the dog. He/she repeats some order while keeping the front legs of the dog in the wanted position. The emotional excitation due to the presence of the master and the hearing of the order activate a link between the aural signal and the body position representation of the poodle. Now, the child repeats the order without keeping the legs of the dog. The wanted body position is now supported by the firing of some virtual signals. This imaginary output is different from the real one. The emotional excitation and this difference activate an impatience signal, which increases progressively the muscular order, therefore the real position (Appendix I). This trial and error correcting process generates a slow answer to the master order. Coincidence between the real and the imaginary positions induces a pleasure impulse, which activates a new link between the order of the master and the convenient muscular order. Bolting of the link is obtained. From this time, the master order induces a fast muscular answer.

Note that the training of a performing dog leads to the building up of a new impatience (induced by the order of the master), a new end signal (reaching the good position), a new reward (the applauses) and a new behavior (the body position): an utterly new instinct has been created.

4.2.5.5.3. Learning to Drive a Car

In the same way, men learn some new body position, for instance any form of salute (elevated fist, hand on the heart or against the kepi). But learning to drive a car confronts us with a new problem: we have not to learn a position, but a gain (the convenient proportionality between the change of direction of the road and the turning of the wheel). Like in the precedent case, a specific label acts as a drive to allow the choice of the driving behavior. Links between the real or imaginative body position (the angle of the wheel) and the muscular orders, therefore on the car position against the road direction, are activated. Trials lead to a slow answer of the learner. Direct activation of the muscular order leads to a much more fast answer. We show (Appendix J) that it can be easily obtained by learning, in place of a position, the amplitude of modulating signal.

4.2.6. The Inputs-Outputs System (A Summary of the Above Section)

Main results: Moto-sensorial boxes are in charge of motion and thinking.

4.2.6.1. Data Flow Through a Motor Controlling System

A typical memorizing and motor controlling device (for instance the device devoted to the sequence of phonemes making a word) includes (Fig. **4.12**):

- Detailed real motor outputs (ordering to utter some phoneme).

- Detailed real specific sensorial inputs (excited when a specific phoneme is heard). If the phoneme is heard just after being ordered, the signal "Me personally" is excited.

- Detailed virtual motor outputs and imaginary sensorial inputs linked through a loop build by a learning process. These virtual inputs and outputs are used to forecast what would be heard after uttering some sound. This loop is used for trial and error correction processing. It can be bypassed. Imaginary sensorial inputs generate a "not real" yes or no signal.

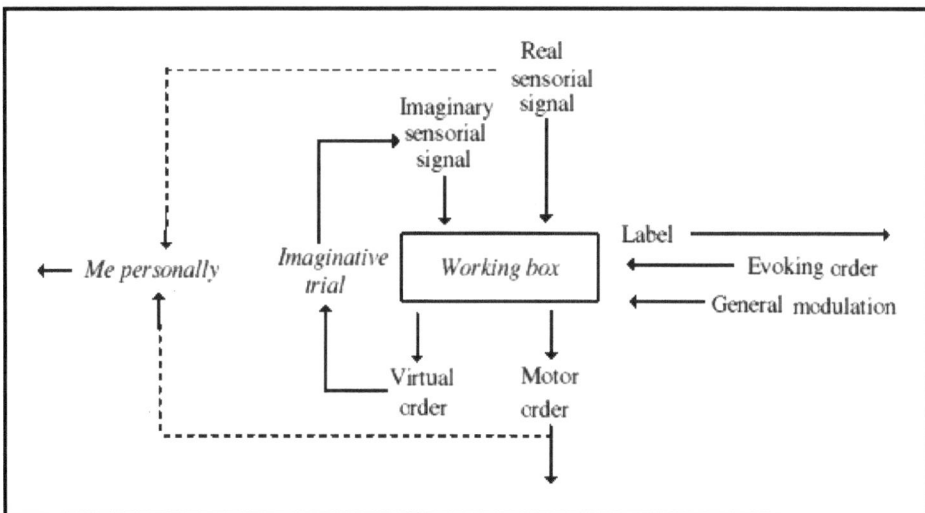

Fig. (4.12). Scheme of a moto-sensorial box.

-       A general modulating activation (often linked to emotions).

-       An evoking input, specific to the whole pathway (specific of the word).

-       An output, which is specific to the whole pathway and which is the label of the word. Transforming the detailed signals into a label signal is a kind of data compressing.

In such hairpin architecture, the evoking signal calls for the sequential excitation of the real or virtual motor outputs, then for the excitation of the label. The firing of the label is an end signal for the system. It often triggers another system and/or emotional feelings.

4.2.6.2. The Main Input-Output Pairs

4.2.6.2.1. The Main Motosensorial Pairs

This discussion applies to the pair speak and hear and to the pair associating body motion and visual detection of body posture. It does not apply to the sense of smell: smells can be felt, evoked and used as labels, but they do not order any motor output.

4.2.6.2.2. The Emotions

They own a triple nature. a) They act in an observable way (widening of pupils, erection of hairs) on the autonomic nervous system); b) they create very strong modulations; c) acting as sensorial signals, they perform a social function.

The Modulating Role of Emotions: From the observation of the faces of various subjects, we deduct that each emotion seems to be either impulsive (short duration, great intensity) or static (several minutes, mean intensity): one easily makes a distinction between delight and well-being, between rage and indignation, fear and light anxiety, between acute pain (just after the death of your mother) and simple static sorrow. Acute emotions or often-repeated static ones will favor recording (activation of new links) while unrepeated static emotion are too weak signals to trigger recording, but they allow the reading of already activated links.

Among the emotions, we give a special attention to pleasure, resentment and pain. Pleasure and resentment are strong signals, triggered by the end of any instinctive drive. They are the effects of reward, disappointment and punishment. We have seen that they govern association and differentiation; we will see that they govern decision.

The Social Function of Emotions: The facial expression of other people is recognized in an inborn way by the right somato-sensory cortex [16, 17]. Sensorial messages from the facial muscles, orders toward the facial muscles and recognizing signals from other people expression combine as a typical example of imitative motor learning (Section 4.5). And, in fact, newborn infants imitate adult facial gestures [18]. We have seen that an output of such a triple device supports an imaginary signal, which acts on the autonomous nervous system as a real one [19] and tends to make feel by the subject the emotions he observes on the faces of other people.

4.2.6.3. Data Flow Through an Associative Memory

Abstractions learning circuits (Fig. **4.9**) are often associated with loop exciting backward neurons (Fig. **4.13**). In this figure, the learning circuit is represented by a white rectangle). Each input is linked to a backward neuron. When the sum of the input excitations is sufficient to excite the common label, <u>all</u> the backward neurons fire (Fig. **4.13** hereafter). Thus, the firing of one of the convergent pathways (supporting for instance an odour) causes the evoking of all the pathways. More details in Appendix H.

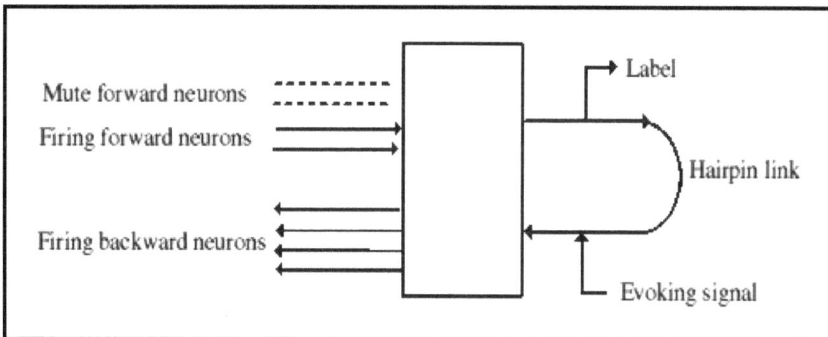

Fig. (4.13). Forward-backward loop.

4.2.6.4. Where in the Brain?

Primary sensorial and motor cortical areas (25% of the cortex) are devoted to inborn analysis of the signals while the mono and multi sensorial associative areas

(75% of the cortex) are devoted to short-term memory and to the formation of long duration associative links. We know for instance (by comparing the Positron Emission Tomography scans for several patients suffering from a stroke [20]) that the memory of people names, of animal names and of tools names are stored in three close places of the left temporal cortex. Yet, memory is not located only in the cortex, but also in deeper structures which support the formation of persistent links (this is probably the case for the cerebellum) or which build up the many modulating signals governing recording and remembering in other places: the reticular formation of the brain stem modulates the thalamus, which modulates the cortex (allowing or not remembrance); the emotional brain (hippocampus, hypothalamus and septum) creates general modulating signals associated with anxiety, fear or rage; the septum is probably in charge of the pleasure signals. Impairments of the hippocampus, close to the septum, (or of only its ventral part [21]) forbid the acquisition of long-term memory, but also of working and even of short-term memory [22]. So, we are led to think that the hippocampus is in charge of the general emotional modulation (which is necessary to learn, not to read memories).

In summary: structures associated with short term and long term memories fill a great place everywhere in the brain.

APPENDICES TO SECTION 4.2

A: Control of short-time memory

B: A repeating device

C: Competition between old and fresh remembrances

D: Bolting the links of a simple reflex

E: Noxious stimuli and weak reward signals

F: Memorizing a melody

G: Bolting the links of a memorizing circuitry

H: Associative memory

I: Trial and error correcting devices

J: Learning the gain of a feedback device

Appendix A: Control of Short-Term Memory

Its main component is a periodic neuron (Fig. **4.14**) controlled by two modulations: a recording one and a reading one.

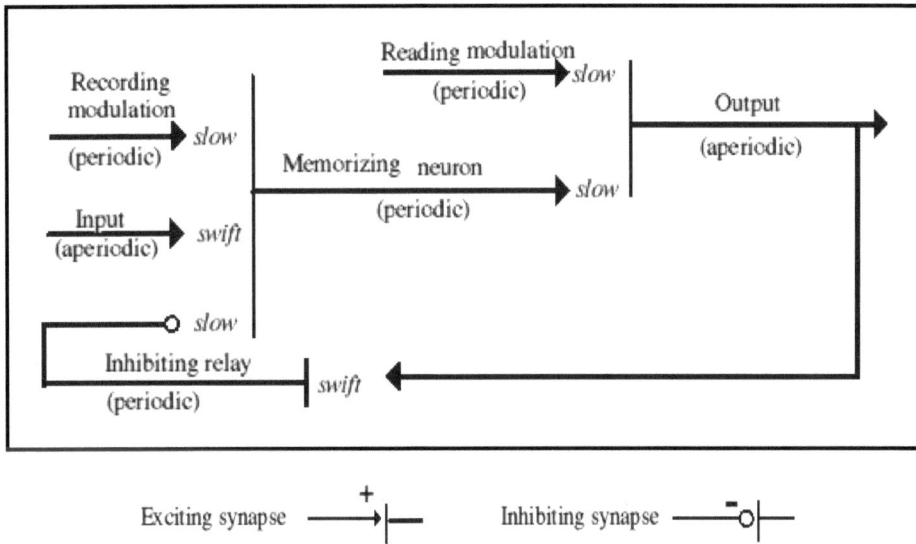

Fig. (4.14). A short-term memory.

In the initial state, the memorizing neuron is silent. It does not start when only the modulation or only the inputs are firing. But, if both are excited, the memorizing neuron begins to fire. It's a periodic neuron: it generates a salvo.

Coupling of the memorizing neuron is insufficient to fire the output. But if the reading neuron is excited while the memorizing one is firing, the output fires and an inhibiting feedback stops the salvo.

Appendix B: A Repeating Device

This device is made of 4 aperiodic neurons and 2 inverting relays (Fig. 4.15). The device has 2 inputs: a modulating periodic and an aperiodic start signals. Synapses N1/N2 and N3/N4 are slow ($T_A \approx 10$ s), the others are swift.

Their contribution to U_C are U_M and U_S such that $U_{ON} > U_M > U_{OFF}$ and $U_M + U_S > U_{ON}$. When the modulating neuron is silent, the device is silent. When the *modulating neuron is firing and after an impulse of the start neuron, the periodic neuron N1 begins to fire. After a delay about T_A, N2 fires, inhibits N1, excites the output and the N3 N4 system symmetric of the N1 N2 one.* Thus, when the modulation is on, the response to a start signal is an output impulse repeated every 20 s.

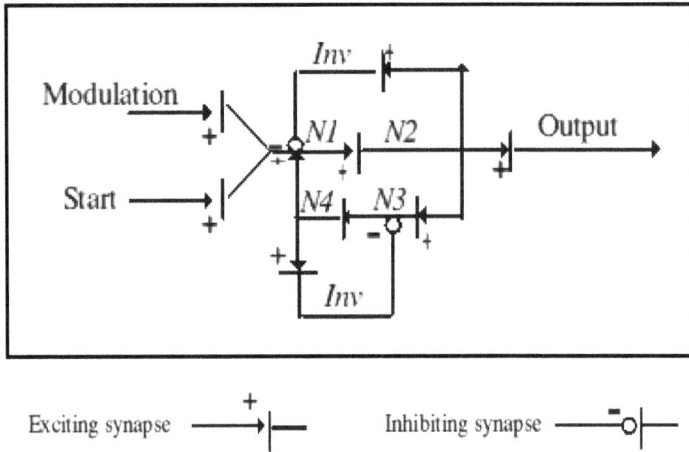

Fig. (4.15). A repeating device.

Appendix C: Competition Between Old and Fresh Remembrances

In Appendix C of Section 4.1, we described learning and reading in a pavlovian link. Now, we look at a set of competitive pavlovian links: the output of each of them inhibits the input of all the others. Each of them has been excited ($n_R = 1$) sometime before. From then, the value of n_R was slowly decreasing. We assume that $F_1 = 0$ and that the firing rate F_2 of the conditional pathway is a slowly increasing modulation, the same for all the neurons of the set. The neuron with the highest n_R value is the first to fire. It inhibits all the others. Thus, a general modulation selects the more excitable link (the more recently excited). Ten minutes later, the n_R of this neuron will increase.

Discrimination of Remembrances

The device selects the more recently excited input. But, due to the noise, it is not always able to discriminate between almost simultaneous inputs. What interval Δt between events can be discriminated after a time T? *If U_{Abs} is known with an accuracy $\delta U = \pm 0.15$ mV, two remembrances with two n_R values can be discriminated) if $\delta n_R > \delta U/8.6 \approx 0.02$ (the more activated output neuron fires first* *and inhibits its competitor. Now, look at the remembrance of two events leading to $n_R = 1$ occurring the first at a date d_1, the second at a date $d_2 = d_1 - \Delta t$. Later, at time $d_1 + T$, we search for old remembrances. The values of n_R are then n_{R1} and n_{R2} (these values are easily evaluated by using Table 4.1. The searching device is able to select the most recent remembrance if $n_{R1} - n_{R2} > 0.02$. (If $n_{R1} - n_{R2} < 0.02$, one either of them is selected by chance).* The following Table **4.3** gives the interval Δt between events which can be discriminated after a waiting time T.

Table 4.3. Discrimination of old remembrances.

Interval Δt	6 Hours	4 Days	6 Years
Waiting time T	24 h	15 d	20 y

For instance, this device does not make any distinction between events occurred 14 and 20 years ago: chronological classification of old remembrances fades away.

Appendix D: Bolting the Links of a Simple Reflex

The Bolting Rules

We deal with n input neurons, some of them firing simultaneously. The n outputs of the first choosing box are neurons Σ_1, Σ_2 ... which cannot fire simultaneously. The intermediate sub-unit owns N columns L_A, L_B. Inside the second choosing box, each of the Σ neurons sends synapses to all the N outputs of the box. Competitive inhibition avoids that several L fire simultaneously. Thus, in a first step, the device chooses some $\Sigma_1 L_A$ association and links it.

The goal of bolting is to avoid that later excitations destroy the specificity of the $\Sigma_1 L_A$ association: ulterior $\Sigma_1 L_B$ or $\Sigma_2 L_A$ associations are forbidden. But $\Sigma_2 L_B$ is allowed: the device can register N different pairs ΣL.

A Possible Scheme

Each Σ neuron sends N synapses to the conditional inputs of N new pathway devices (so, there are nN such devices in the sub-unit). The L neurons are the conditioning inputs. Once such a pathway (for instance $\Sigma_1 L_A$) is opened, the firing of Σ excites directly the column L_A and inhibits all the L entrances in the second box (so forbidding any $\Sigma_1 L_B$ linking). On the other hand, the general Σ signal (Σ_1

OR Σ_2 OR Σ_3 ...) *acting as a conditional input and* L_A *as an conditioning one open a new pathway, which inhibits the* L_A *circuitry inside the second box: thus, links such as* $\Sigma_2 L_A$ *are forbidden.*

Appendix E: Noxious Stimuli and Weak Reward Signals

Noxious Signals

We have seen that the learned link is not created at the conditioning time, but some ten minutes later. Thus we expect that the activation of a sequential chain of n links need n successive repetitions. But we observe that a short noxious stimulus is sufficient to induce the at once activation of several sequential links. This means that in this special case the stimulus does not act directly on the learning device: some periodic neurons, triggered by the short noxious signal, carry on firing for more than ten minutes, allowing the links to be successively activated after only one stimulus.

Weak Rewards

The molecular mechanism of NMDA receptors explains also the occurrence of slow activations: before learning, the firing of the output of a three neurons device depends on the frequency of the conditioning input. We have seen that, during each spike, the NMDA (activating) glutamate receptors let some Ca ions enter into the cell. These ions trigger an increase of the number of activated AMPA/kaïnates (main) glutamate receptors. Thus, the conditional synapse becomes operational after a fixed number of conditioning spikes. Weak rewards provide only a weak increase of the activation of the main receptors; a great number of repetitions are needed.

Appendix F: Memorizing a Melody

We look here only to the evoking device and to the virtual output.

The Backward Neurons

N backward neurons B (one for each of the labels) make a hairpin structure. They go from the upper floor toward the lower floor. They can be triggered either by an autonomous evoking signal or by the combination of the melody output label AND an emotional modulation (thus ordering to repeat the evoking).

*Each B neuron goes through all the floors. Fig. (**4.16**) shows its links within a floor k. B is used as the conditional input of a NP; its conditioning input is the combination Lα AND Em; its output is Dα (N such new pathways for each of the N neurons B in each floor: these links ensure the specificity of the evoked melody). On the other hand, let Λ be a neuron excited (through OR operations) by any neuron of the floor (k-1). A neuron Λ*$_D$ *(D for delayed) fires a time Δt after Λ. Dα AND Λ*$_D$ *excite Lα. (In the first floor, Dα alone excites Lα).*

During recording, new pathways are opened. After recording, if an evoking signal triggers the neuron B, the signal propagates backward, the convenient Dα of each floor are excited. The B signal reaches the first floor, causes the first floor Lα to fire. After a delay Δt, the Lα of the second floor fires: once the recollection has been initiated, it spreads toward the second, then to the third floor. In summary, the function of Dα is to preserve the specificity of the theme and the function of Λ is to maintain the orderly sequencing of the sounds.

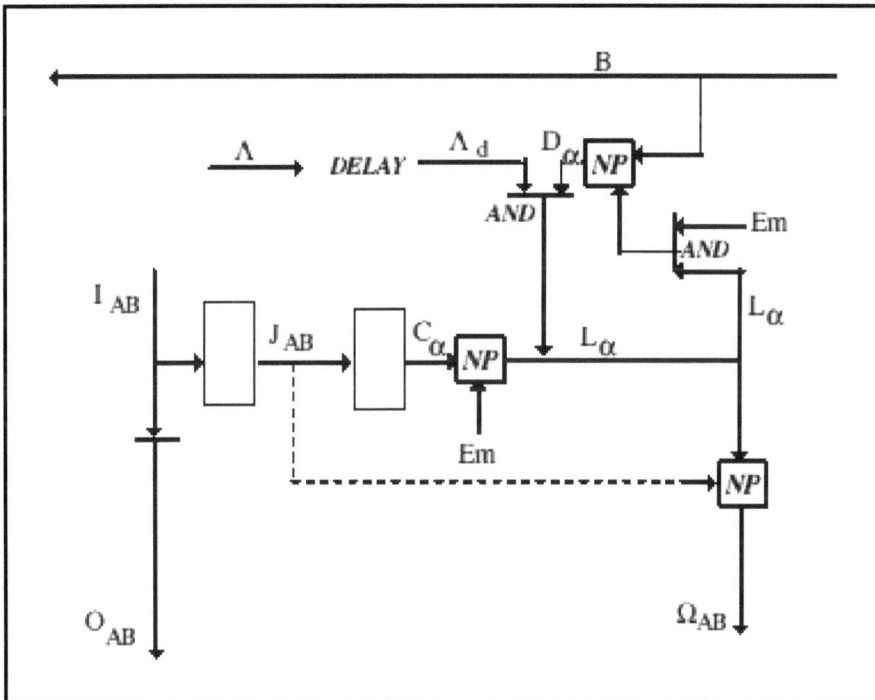

Fig. (4.16). Part of backward neurons for memorizing a melody.

Recording the Outputs

Lα is used as the conditional input of a second NP; J_{AB} (dotted lines on the figure) is used as the conditioning input (nN such devices in each floor). The output is the **virtual sensorial output** *Ω_{AB}. Remind of the main property of NP devices: after recording, their output fires only if the conditional input is firing and the conditioning one is mute. Then, after recording, Ω_{AB} is mute if J_{AB} fires; Ω_{AB} fires if Lα fires while the real input J_{AB} is mute, that is only during evoking.*

Appendix G: Bolting the Links of a Memory Circuitry

Bolting rules for a sequential memory are slightly different (except for the first floor) from the rules for simple reflexes because the same note can be repeated several times.

The Bolting Rules

1)   A rhyme made of the notes ABC cannot have two labels.

2)   The rhymes ABC and ABD cannot have the same label.

3)   The rhymes ABC and DBC cannot have the same label.

Transposed Rules

From the above rules, we extract the rules for a floor (for instance the second floor).

After recording of the theme ABC, the sensorial signal J_1 (an abbreviation for J_{AB}) was linked with the column α of the first floor, J_2 (for J_{BC}) with the column β of the second floor and fast direct lines (bypassing the second choosing boxes) linking J_1 to Lα and J_2 to Lβ have been opened.

1)   If the second rhyme begins by the same J_1 (and thus is linked to the same Lα), rules for the second floor are: J_2 is linked to the same β (J_2Lγ is forbidden). The column β cannot be associated with a signal G_2 different from J_2: G_2Lβ is forbidden. (But G_2Lγ is allowed).

2)   If the second rhyme begins by H_1 different from J_1: J_2Lβ is forbidden. (But J_2Lγ is allowed).

A Possible Scheme

If the second rhyme begins by the same J_1, the rules for the second floor are the same as the bolting rules of Appendix D and the same circuitry can be used. The problem for the second floor is to detect a change of the input of the first floor (H_1 in place of J_1). The link between the columns of floors 1 and 2 should be in charge of N^2 new pathway devices. The delayed signals $L\alpha$ of the first floor act as conditional signals, the signal $L\beta$ of the second floor as an conditioning signal to open a new pathway $P\alpha(\beta)$. A feedback inhibition avoids that, later, α' links with β. So, β is definitively associated to α, but α can be associated with several β. The neuron $P\alpha(\beta)$ excites or inhibits the bolting device described in Appendix D.

Appendix H: Associative Memory

The scheme of a forward associative device (Fig. **4.5**) is completed by backward neurons and new pathway links (Fig. **4.17**). The firing of the label triggers the firing of all the linked backward pathways.

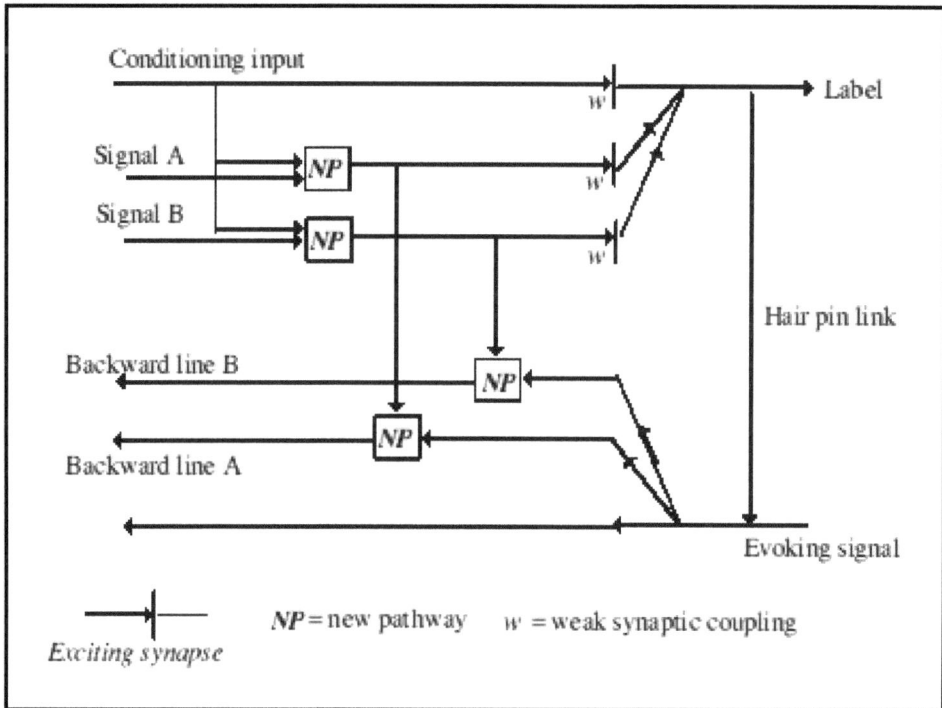

Fig. (4.17). New pathways and hair pin loop in an abstractions making memory.

Appendix I: Trial and Error Correcting Devices

We look at the circuitry which tests a pre-motor order, modifies it till being sure it will give the expected motion and only then lets it govern the muscle (Fig. **4.18**).

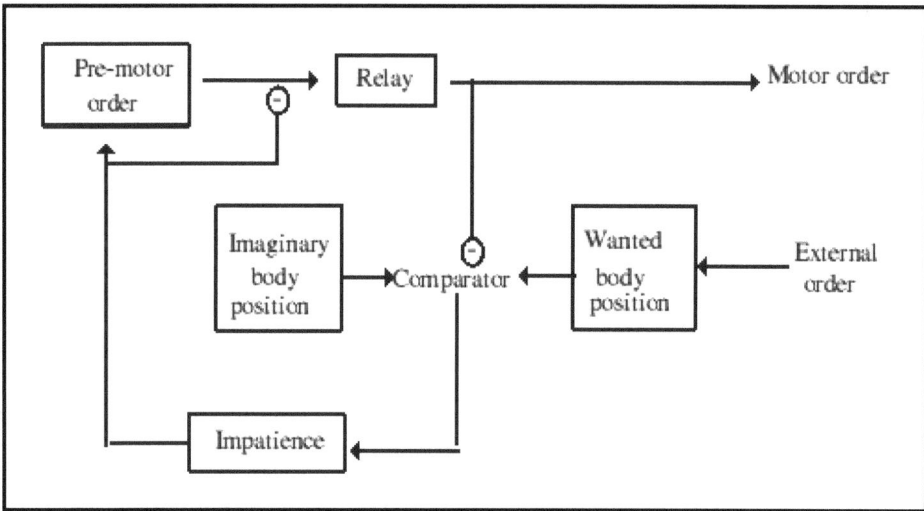

Fig. (4.18). Imaginary preparation to a body motion.

The external order (the voice of the dog' master) induces a wanted body position signal. The imaginary body position signal is an output of the pre-motor box. If smaller than the wanted position signal, it inhibits the relay and it induces an increase of the pre-motor order (we called this last signal impatience). When the imaginary position reaches its wanted value, the inhibition of the relay stops; a signal is transmitted to the muscles and the trial and error correcting device is inhibited.

Appendix J: Learning the Gain of a Feedback Device

Learning to drive a car requires the learning of the convenient gain G of the controlling part of the feedback linking the car orientation to the wheel position: for the simplest control system, wheel position correction $= G *$ (wanted orientation – observed orientation). Thus, we would have to control by a multiplicative factor the muscular order and we do not know how to build up multiplicative wirings with neurons.

*By luck, we remember that muscular orders command what we called the muscular excitation Φ, which is in fact the ratio of recruited contractile fibres. But the orders are transmitted through motoneurons and we have seen in 3.3 that Φ is an exponential function of the ratio **k** of recruited motoneurons. So, to multiply Φ by G, we have only to add to **k** a constant term Log(G). Adding a constant term is easily.*

REFERENCES

[1]   Morris RG, Anderson E, Lynch GS, Baudry M. Selective impairment of learning and blockade of long-term potentiation by an N-methyl-D-aspartate receptor antagonist, AP5. Nature 1986; 319(6056): 774-6.

[2]   McHugh TJ, Blum KI, Tsien JZ, Tonegawa S, Wilson MA. Impaired hippocampal representation of space in CA1-specific NMDAR1 knockout mice. Cell 1996; 87(7): 1339-49.

[3]   Philpot BD, Weisberg MP, Ramos MS, *et al*. Effect of transgenic overexpression of NR2B on NMDA receptor function and synaptic plasticity in visual cortex. Neuropharmacology 2001; 41(6): 762-70.

[4]   Malenka RC, Bera MF. LTP and LTD: an embarrassement of riches. Neuron 2004; 44(1): 5-21. Review.

[5]   Sibille EL, Pavlides C, Benke D, Toth M. Genetic inactivation of the Serotonin(1A) receptor in mice results in downregulation of major GABA(A) receptor alpha subunits, reduction of GABA(A) receptor binding, and benzodiazepine-resistant anxiety. J Neurosci 2000; 20(8): 2758-65.

[6]   Klein M, Kandel ER. Presynaptic modulation of voltage-dependant Ca2+ current; mechanism for behavioral sensitization in Aplysia californica. Proc. Natl Acad Sci USA.1978; 75(7): 3512-6.

[7]   Changeux JP, Courrège P, Danchin A. A theory of the epigenesist of neuronal networks by selective stabilization of synapses. Proc Natl Acad Sci USA 1973; 70(10): 2974-8.

[8]   Lorenz K. Der kumpan in der umvelt des vogels. J für Ornithol 1935; 83: 137-213.

[9]   Preuss TM, Goldman-Rakic PS. Crossed corticothalamic and thalamocortical connections of macaque prefrontal cortex. J Comp Neurol 1987; 257(2): 269-81.

[10]   Menzel R. Behavioral access to short-term memory in bees. Nature 1979; 281(5730) : 368-9.

[11]   Rakic P. Neurogenesis in adulu primate neocortex: an evaluation of the evidence. Nat Rev Neurosci 2002; 3(1) : 65-71. Review.

[12]   Pavlov IP. Typologie et Pathologie de l'Activité Nerveuse supérieure. P.U.F. Ed.. Paris, France 1955.

[13]   Dehaene S, Naccache L. Towards a cognitive neuroscience of consciousness: basic evidenceand a workspace framework. Cognition 2001; 79(1-2): 1-37.

[14]   Courtney SM, Ungerleider LG, Keil K, HaxbyJV. Transient and sustained activity in a distributed neural system for human working memory. Nature 1997; 386(6625): 608-11.

[15]   Kritzer MF, Innis RB, Goldman-Rakic PS. Regional distribution of cholecystokinin receptors in primate cerebral cortex determined by *in vitro* receptor autoradiography. J Comp Neurol 1987; 263(3): 418-35.

[16]   Adolphs R, Damasio H, Tranel D, Cooper G, Damasio AR. A role for somatosensory cortices in the visual recognition of emotion as revealed by three-dimensional lesion maping. J Neurosci 2000; 20(7): 2683-90.

[17]   Iacoboni M, Woods RP, Brass M, Bekkering H, Mazziotta JC, Rizzolatti G. Cortical mechanisms of human imitation. Science 1999; 286(5449): 25268.

[18]   Meltzoff AN, Moore MK. Newborn infants imitate adult facial gesture. Child Dev 1983; 54(3): 702-9.

[19]   Levenson RW, Ekman P, Friesen WV. Voluntary facial action generates emotion-specific autonomic nervous system activity. Psychophysiology 1990; 27(4): 363-84.

[20]   Harvard Medical School: News Release (on Internet) August 2006

[21]   Maggio N, Segal M. Unique regulation of long term potentiation along the septotemporal axis of the hippocampus. Hippocampus 2007; 17(1):10-25.

[22]   Hartley T, Bird CM, Chan D, *et al*. The hippocampus is required for short-term topographical memory in humans. Hippocampus 2007; 17(1): 34-48.

CHAPTER 5

The Thinking Brain

Abstract: Social instincts need an exchange of messages. In animals, real or imaginary heard signals act as instinctive drives to trigger a behavioral response (flight, attack). In men, the labels of real or imaginary heard sequences of phonemes are words and words act as instinctive drives to utter other words. Real or imaginary speeches allow long term strategies.

A man can learn and recognize about 40 phonemes and less than 10^4 words (when modern dictionaries quote 10^5 words). The output of a word device is used as a partial instinctive drive. The total drives result from a fluctuating mosaic of partial drives, a word calls for another word. Sentences are seldom stored in a long term memory, but compacted in a short-term memory. Primary words are related to motor behaviors. Others are abstract concepts related to primary words or abstractions built from abstractions. They allow the bottom-up building of long term strategies. Their results are evaluated by downward imaginative tests. The frontal cortex confronts the immediately most pleasant behavior (insulting my boss) and the long term most pleasant one, explaining how works self-control. Mechanisms of language explain fiction, creativity and rational thinking. In dog as in man, another complex moto-sensorial device governs complex motions. Simple motions play the same part that phonemes, simple actions replace words and complex actions replace sentences. Combining language and complex motions, man becomes able to conceive new actions, to invent and use tools. Human consciousness is possibly related to the imaginative working of these two moto-sensorial devices. Then, dog's consciousness would be related to its imaginative motions.

Keywords: Abstract concepts, action, complex motions, consciousness, cortex, creativity, fiction, imaginative trials, imagination, language, phonemes, rational thinking, self awareness, self-control, sentences, social instincts, strategies, thought, tools, words.

5.1. The Word

5.1.1. The Road Toward Intelligence

Main facts: Intelligence, which allows long term goals, uses multi drives architecture.

5.1.1.1. The Social Instincts

The one-neuron sea anemone ejects seeds in the water as plants eject pollen in the wind. In more advanced species, the outcome of using intromission leads to a huge economy of seeds, but requires the exchange of signals to recognize and locate possible partners. More generally, social instincts (sexuality, collective hunting, collective defense) need exchange of messages. In butterflies, they are supported by pheromones. In frog, birds, mammals, they are supported by sight, smell and earing.

Look at a raven on sentry duty. After a suspicious noise, the bird caws. Its own shout does not trigger any response. Hearing it, the other birds fly away. This behavior results obviously from a moto-sensorial device (see Fig. **4.12**). In men, hearing, speaking and writing are linked in a specialized cortical area [12].

5.1.1.2. Long Term Goals and Strategies

5.1.1.2.1. Short Term Goals

Motions triggered by the exploring instinct do not have any goal: without any external stimulus, primates investigate the strange gadgets placed in their cage; random walk is used by rats to explore a labyrinth without any apparent benefit. Results are recorded in a short-term memory. If some good smelling food is place at the output of the labyrinth, a goal then appears: finding the food. The map of the shortest way that has been recognized is transferred to a working memory [1], the external stimulus triggers the learning of the best way.

Most of animals' complex motions are ruled by immediate goals: when a squirrel is afraid, it has to build up a simple strategy (to detect a tree, to imagine and optimize its run and its jump), then carry on these elementary motions with the immediate goal to hide.

In man, the process begins with the choice of a general goal (for instance to go and buy a newspaper). Starting from this general goal, human beings must define a strategy: it is the list of immediate goals (go to the door, open it, turn right and walk till the newsstand) to be successively reached. (Fig. **5.1** hereafter). If they do not know where the shop is, as it is the case for someone who just moved in to a new city, they undertake a random trials and error correction process. Once he/she has learned where the shop is, a direct line (see 4.2.3) bypasses the trial device.

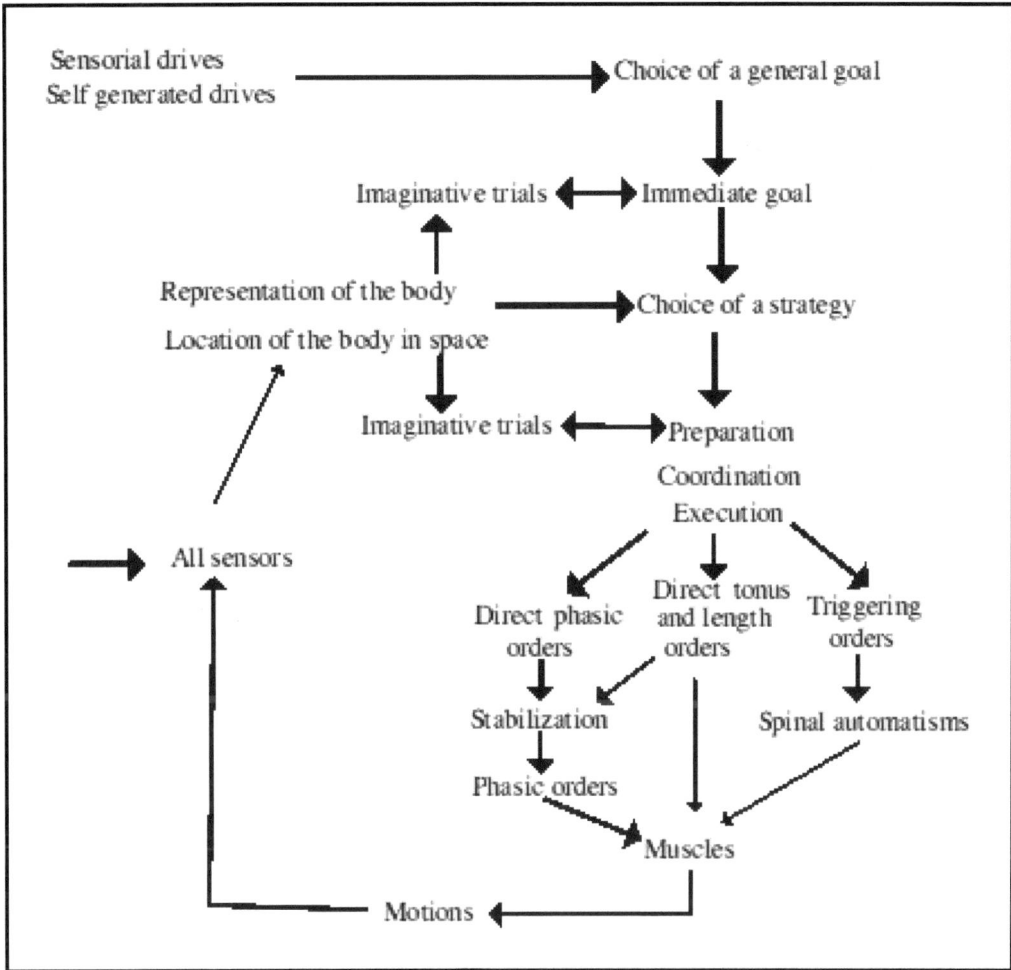

Fig. (5.1). Block-diagram for the choice of a strategy.

5.1.1.2.2. Long-Term Goals

Men are not only moved by short-term goals, but also long-term goals. Ambitious or malignant people (such as Shakespeare's Iago) design far-fetched strategies clearly supported by a great number of intermediate choices.

5.1.1.2.3. Tools

Animals do not exhibit long-term goals, maybe because we are unable to observe them, except perhaps elephants: is their long walk through the savanna to find water driven by a long-term goal?

Conceiving, manufacturing and using tools also require a great number of intermediate steps. Some animals exhibit an embryonic industrial activity: birds use twigs and feathers to manufacture their nest; apes use a stone to open a nut. But these situations only require a little number of successive choices.

5.1.1.2.4. A Possible Quantitative Definition of Intelligence

Human intelligence seems to result from a great number of sequential choices. Could the intelligence of a chess player be quantified by the number of moves he anticipates? In this case, Blue Gene would be more intelligent than Kasparov. But everyone is not, or not only, a chess player. The self-organizing sub-units of the brain have a limited number of columns. Progressive learning leads them to turn to chess, poetry or politics. Thus, one of the many definitions of intelligence, and perhaps a substitute for the IQ, would be the number of degrees of the analyzing pyramid used by each subject.

5.1.1.3. The Architecture of the Areas of the Intelligent Brain

Primates and men use the same neuronal components. The difference between their intellectual performances cannot be explained only by an increase of their neocortex size (7 billion neurons in gorilla, 20 billion neurons in man). We have to search, using self-observation, for some small change in the way the components are connected leading to a qualitative behavioral change.

Our imagination sometimes leads to sensorial and highly emotional evocation of a short duration scene (some childhood recollection). It can also result in an abstract, time compacted, poorly emotional speech (the story of my family since the crusades). The first one is ruled by the normal architecture, the second one by an unusual, special one. In the special architecture, each word calls for another word and the internal speech can work in a perfectly autonomous way. It can also be coupled to real motosensorial signals, emotions and instinctive drives and more generally to the normal architecture. Thus, in man, this highly performing architecture is put upon the top of a common one, present in many animals.

Any behavior results from a choice or several sequential choices between competitive drives. In most of the cases, the drive signal is the sum of several modulating signals and each of these modulations excites a particular set of several neurons. The chosen neuron is at the intersection of these sets. In the common architecture, the drive is the sum of a little number of strong modulations; the responses follow a rigid pattern. (Procedural memory is probably due to the normal architecture, declarative memory to language and more generally to the special architecture). On the opposite, the responses of the special architecture follow a remarkably fluctuating pattern: the drive must be the sum of a large number of weak modulations.

An abstract reasoning is a sequence of dynamic periods (the uttering of some sentences) separated by static intermediate results, which can be evoked by or combined with any other intermediate result: the inputs and outputs of the special architecture must be linked to motosensorial pairs. The most easily usable is the hearing and speaking pair, the labels of which are phonemes and words. You may also use the viewing and hand moving pair, the labels of which are sketches and written letters. Thus, intelligent activity is strongly associated with language and drawing, which allows tools. We will now describe the language mechanism.

5.1.2. What Everybody Knows About Languages

Main facts: How babies pass from proto-language to adult language.

5.1.2.1. Oral Language Formation in Men

5.1.2.1.1. Crying

The many paths of the complex moto-sensorial device are successively linked. The first step is crying. Newly born babies weep and cry. The airflow coming from the lungs goes through the larynx, which is partly obstructed and separated from the pharynx by the vocal cords. The pharynx emerges into the mouth and the nasal fossae. Several muscles act to modify the tension and the position of the vocal cords, the position of the tongue, the jaws and the lips, the communication between the pharynx and the nasal fossae.

Crying seems to be an inborn automatism triggered by pain or wrath. The excitations of each of the muscles acting in phonation reach there per defect value. For instance, the crying strength (the lung acting) is at its maximum.

5.1.2.1.2. Babbling

When babies are 2 months old, an exploring instinct leads them to explore their motor functions. (While the green monkey (Cercopithecus Aethiopis) babies exhibit exploratory shouting and babbling, our supposedly closest cousins, the chimpanzees, which own a rudimentary rhino-pharynx, are reported to suffer of an absence of exploratory phonation). Babbling is the random exploration of the phonetic motor apparatus. Self-hearing creates a pleasure signal (a reward), which will be used in the next step to activate new reflexes. The inborn crying instinct is gradually inhibited by the babbling drive: babies cry every night, children rarely cry, adults never cry. Quantitative control devices of all the muscles of the phonetic apparatus (governing the lungs, the throat, the lips, the jaws and the tongue) are activated: the babbling strength is not excessive.

5.1.2.1.3. The Proto-Language

A second step occurs about 9 months after a baby's birth: the search for social communications. Modulated shouts, facial expression and gestures are used as a proto-language, obviously due to the common architecture. The signs are not combined; each sign acts as the drive of some basic instinct such as fear, hunger, and aggressiveness. End of drive occurs when the parents pay attention to the child and give some specific response.

At the same time, as a result of babbling, links are created between emitting and hearing phonemes, which are not yet used for communication. When babies are 12 months old, they insert some words (less than 20) in their proto-language.

5.1.2.1.4. The Beginning of Adult Language

When they are about 15 months old, babies replace the proto-language by a purely verbal one. This change corresponds to the implementation of the specific architecture proper to human being. Babbling disappears. Its drive is replaced (or masked) by the need for social communications. The number of words learned each month was of 5 before this event, then of zero during a three-month transition after which babies learn about 20 words each month.

At first, babies use a language made of some very simple words to act on his parents. For instance, "milk" can either mean "I see the milk bottle" or "I want some milk". The success (a reward) accelerates the learning process. Then appear two-word sentences made of a subject and a verb, for instance "baby eats". When children are 2 years old, they use the pronouns I and you. At 3 years old, they own a 300-word vocabulary and begin to use past time and plural. A new random exploring period appears when children are 4 years old: they are constantly speaking and asking why. When they are 5 years old, some children use a 2,000-word vocabulary; they master the syntax and show signs of self-criticism. It is time for them to learn how to read. At the same age, other children only use a 200-word vocabulary.

5.1.2.1.5. A Growing Vocabulary

Children live within a flow of words. The talking in their family acts as a random source of unknown words. Children learn the ones they need for their actual development. A new link is activated between the hearing of some word and the simultaneous excitation of a newly appeared sensorial signal, generated from the older ones by differentiation and association. Linguist of the structuralist movement speak of distinctive and combinative functions. Thus, at first, children associate themselves with the name or the nickname used by their parents to call them. Later, differentiation leads them to separate "I" (themselves as felt by themselves) from their nickname (themselves as seen by other people) and from "you" (the person speaking to them). The pronoun "we" is introduced as the association of I and you, and the pronoun "they" as the Boolean negative of "we".

Later, new words are associated not only with primary sensorial invariant signals, but also with more abstracts inputs related to the body position, behavioral drives, emotions, pre-motor orders and, in a last step, to very abstract signals.

Thus, the number of words learned is continuously increasing at least during some years: most of the people use less than 3,000 words and some only 200.

5.1.2.1.6. Utilitarian, Extrovert and Inner Languages

When language replaces proto-language, utilitarian drives, governed by need of attention, of help, of food, inhibit the exploring drive.

When children are about 4 years old, a variant of exploring drive appears, which acts as a general exciting modulation: children are continuously speaking and asking "why?". The answers are useful because they increase the number of registered labels, even if the rational meaning of the explanations is utterly lost.

This exploring instinct does not disappear with age: the need to speak is obvious in teenagers and in many adults.

In most people, a new instinct appears some time after the extrovert language. It favors imaginary hearing to the detriment of real speech. Humans hold then a dialogue with themselves, reaching the realm of thoughts.

5.1.2.2. Gestured and Visual Languages in Man

5.1.2.2.1. Inborn Proto-Languages

Other kinds of languages use the pair vision and posture or the pair vision and body motion. Some of these languages, obviously ruled by the common architecture, are inborn: someone's emotive status (the mood) is expressed in their physiognomy and any other man understands the message. Grimaces are a universal language. Everybody understands gestures meaning "come here", "look there" or "go away".

5.1.2.2.2. Learned Proto-Languages, Mnemonic Objects

A Pavlov link associates some arbitrary stimulus (the view of an object) with the drive triggering a behavior or an evocation. For instance, you can make a knot in you handkerchief to remember something. Such tricks have been used by ancient peoples of Eastern Europe: Scythes and Germans exchanged notched sticks, each notch having a mnemonic meaning.

5.1.2.2.3. Drawings

The special drawing system does not use the general pairs vision and general posture or vision and body motion; it uses the specialized pair vision and hand motions to represent the visual environment and to express intermediate results for abstract reasoning and for manufacturing tools. It uses the same pair to carry out writing.

Drawing is the simplest of these languages: associative devices respond to simplified sketches by evoking the whole picture. Speech is convenient to build abstractions, but not to describe faces or landscapes. On the contrary, drawing is convenient to describe faces or landscape and to represent paths between abstractions (see for instance the Feynmann's diagrams). Drawings appeared centuries before any kind of writing. In North America and in some Pacific islands, the memory of famous events was kept thanks to cartoon strips-like sequences of drawings. One example of them is the Bayeux's tapestry (which recalls the conquest of England by the Normans). Everybody understands them, but comment them by using his own words.

5.1.2.2.4. Writing and Reading

Writing has two advantages: a) it is faster than drawing; b) the reader uses the same words as the writer. Writing began in all civilizations (between 4000 and 3000 BC in China, in Mesopotamia and in Egypt, later in America) by pictograms associated with an object (for instance the sun) and the name of this object. Then appeared ideograms, which are symbolic representations associating not only objects, but also abstract ideas (for instance the light) with the sound and the meaning of words. In a third step, signs were associated with part of words to create syllabic hieroglyphs. Such writing required a great number of different signs. The alphabetic signs were invented much later (in about 1500 BC). They link phonemes without any direct association to the meaning of words. Their advantage is that they only require 22 to 26 signs.

Reading is acquired thanks to an imitative process, which leads to the activation of links between hearing, seeing and acting. Alphabetic-reading signals excite the phoneme recognizing devices; ideogram-reading signals excite word-recognizing devices as ideograms are assimilated to words. Some charlatans have imposed a "global method" to learn how to read: they assume in spite of experiment that we use syllabic signs. Pupils must discover by themselves the notion of alphabetic signs, a notion that has historically been so difficult to invent. Such a method cannot work, and experimentally, it does not work.

From the study of some cortical lesions, we know that some localized areas are devoted to reading and writing. But the invention of the alphabet is such a recent event that we are sure that the brain has already contained these areas before the appearance of writing. In illiterate peoples, these areas are probably devoted to another task, perhaps to visual imagination.

5.1.2.2.5. Some Variants

Typed Letters: Another hand motion (typing in place of writing) is associated with reading.

Secret Messages: Only initiated peoples can decode it. In some civilizations, they were used before the invention of writing. For instance, in Peru, the quippos (diversely colored wool cords, each of them with a knot at a variable height) were the only writing known at the Inca period.

Deaf-and-Dumb Language: Deaf-and-dumb peoples give an example of learned visual language. In their cortex, the structures usually devoted to sound analysis are not activated. Deaf-and-dumb language expands into these void areas: the proof is that lesions inducing some kind of aphasia in normal peoples induce the corresponding failure of expression or understanding in patients using the deaf-and-dumb language.

5.1.2.3. Some Examples of Animal Languages

Most of animals express their emotions [2]. If we call "language" the support of this expression, all animals have a language. By some aspects, the languages of birds or mammals are similar to human proto-language.

The language of crows is made of a little number of signals (for instance an alert signal), each of them associated to a behavior. A visual alert triggers the sentry shouting. Other crows hear it. If the loop ordered shout-heard shout excites the signal "Me personally", the bird continues to watch. If not, it takes flight.

In parrots, a motosensorial pair governs the hearing and the uttering of phonemes. Although the brains of men and parrots are very different, their phonemes-recognizing devices seem very similar and Broca's aphasia in parrots has been described [3]. But in parrots (as in some bad pupils), phonemes excite directly some sequential memory: the birds repeat without understanding. In man, on the opposite, each word owns a meaning.

Lycaons (carnivorous connected to both dogs and hyenas) own a proto-language. When hunting big preys, they use of a group strategy with frequent changes of the role of each attacker. This strategy is continuously coordinated by an exchange of barking.

The green monkey (Cercopithecus Aethiopis, known for its blue testes) uses a proto-language composed of 60 gestures and postures and 35 shouts [4] (almost one hundred signs, while the usual vocabulary of a poorly intellectual man is about 200 words). This language is not fully inborn: children monkeys make errors and the meaning of a shout is different from a tribe to another.

5.1.2.4. From Monkeys to Men

Cortical areas are three times larger in men than in monkeys. The additional neurons increase the associative areas, allowing language. We have described

(Fig. **4.13**) the abstraction making circuitry: several detailed lines are linked with a general line. The device can be read forward and backward. In animals, really or imaginary heard signals act as instinctive drives to trigger a behavioral response (flight, attack). In men, some words act (as in animals) as drives of some behavioral response and in the same time as detailed lines of some abstractions. Most of words act simultaneously as detailed lines and general lines of some abstractions, allowing a pyramidal reading process (up, toward more and more abstract messages and down, toward motor behaviors), thus allowing long term strategies.

5.1.3. Phonemes and Words

Main results: How to build meaningful signals.

5.1.3.1. The Phonemes Recognizing Device

Words are sequences of phonemes and sentences are sequences of words. But there are some basic differences: phonemes are learned once and for all and do not have any meaning; words are learned once and for all and have a precise meaning; sentences (with the rare exception of learned-by-heart texts) are quickly forgotten. In this Section, we will describe the learning and uttering of the phonemes and the words. In the next Section, we will study how words build up sentences and how sentences call for words.

5.1.3.1.1. The Spectral Properties of Phonemes

The airflow through the partly obstructed larynx is unsteady: it produces a sequence of air pressure puffs (40 to 100 puffs per second) with a very quick uprising and a slower decrease. The acoustic energy W is distributed among a wide continuous spectrum with frequencies F from 40 to 8000 Hz (Fig. **5.2a**). Then, the pharynx, the mouth, the lips and the nasal fossa act as a set of complex resonant cavities: the acoustic energy is now concentrated around the main resonant frequency of the system and some of its harmonics. The pressure exerted by the lungs governs the magnitude of the acoustic energy. The pitch of the voice is driven by the length of the vocal cords, which depends on age and sex, and by their tension. Vowels are almost stationary sounds: their spectrum is made of narrow bands around some harmonics (Fig. **5.2b**). Experiments show that the location of the three lower bands is sufficient to recognize a vowel. Consonants are transient sounds with some high frequency bands (Fig. **5.2c**).

5.1.3.1.2. The Number of Recognized Phonemes

Looking at the 6,500 known languages, one finds a worldwide inventory of about 10^4 used phonemes. But each language uses dramatically less: in French, there are 14 vowels or diphthongs and 22 consonants (a total of 36 phonemes). And there are less in some dialects, for instance Lappish and old Tahitian.

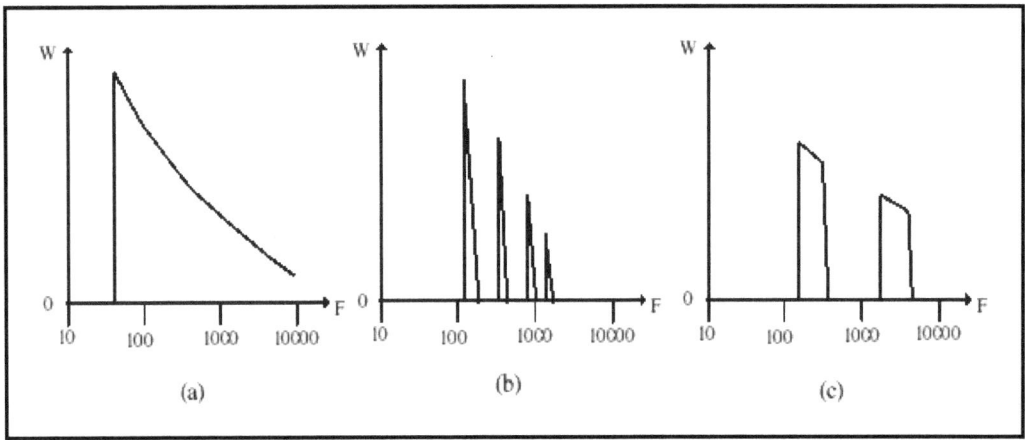

Fig. (5.2). Spectral properties of phonemes. **a)** 40 puffs/s, **b)** pitch 150 Hz- Vowel "O": harmonics 450, 900and 1200 Hz, **c)** pitch 150 Hz- Consonant "J": bands 150 to 300 and 2400 to 6000Hz.

5.1.3.1.3. Criteria for Phonemes Recognition

From one speaker to another, the sounds strength and the pitch of the voices are different. As a result, the primary auditory signals used as inputs of the device must be invariant for sound intensity and for the main frequency of the sound. As for phonemes recognition, the only possible invariant signals are the frequencies of the harmonic bands pertaining to the main frequency and their relative intensities.

The frequencies of the more intensive harmonic bands are used as components of a password (see Section 5.2). Experiments show that only 3 harmonic bands are used. As a result, the device only requires 6 floors (3 relative frequencies and 3 relative intensities) to select a 6-signal password.

5.1.3.1.4. The Device Structure

The working of such a device was summarized in 4.2.6. Its 6-floor can be excited by 10^4 possible phonemes. Each floor receives an excitation from a competitive set of N possible signals. N is such that $N^6 \approx 10^4$. Thus, N = 5. The sub-unit is excited by 6*5 = 30 real aural inputs (the number of possible different invariant signals) and by the same number of imaginary sensorial inputs. On the other hand, we know that English or French contain about 36 phonemes among the 10^4 possible ones: the sub-unit has 36 columns.

5.1.3.1.5. Working of the Phonemes Device

Learning: Coincidence between emotional modulation and hearing of a phoneme causes the learning of a password.

Listening to an External Signal: After learning, the label fires when hearing a phoneme fitting the password.

Learning of the Muscular Command: Babbling allows building up the loop muscular order-aural sensation.

Speaking: Firing of the evoking input when the real motor output is opened induces the uttering of the phoneme.

Imaginative Mode: Firing of the evoking input when the real motor mode is inhibited induces an imaginary hearing.

Links with Reading and Writing: When using alphabetic writing, the visual recognition of letters is linked to the phoneme labels by a device similar to the above ones and the hand motion orders is linked to the letters recognition.

5.1.3.2. The Words Device

5.1.3.2.1. The Number of Words

Words are combinations of phonemes and intonations. In the European languages, words contain less than 10 phonemes. We do not discuss the part of intonations, which play a minor role (for instance to denote anger or interrogation), although in some Asiatic languages (such as the Chinese), intonations play a crucial role while words contain at most 3 phonemes.

The English usual language completed by compound words, technical neologisms and people's and places' names contains an almost infinite number of words (modern dictionaries include from 30,000 to 120,000 words). On the other hand, most people use from 3.10^3 to less than 300 words and only 200 irregular verbs.

But many people know several languages. Hence, we assume that our brain is able to memorize about 10^4 or perhaps 10^5 words. Words are obviously driven by a sequential memory device (Fig. **5.3**). This device contains fewer than 10 floors and some 10^4 or 10^5 columns, probably fragmented into groups.

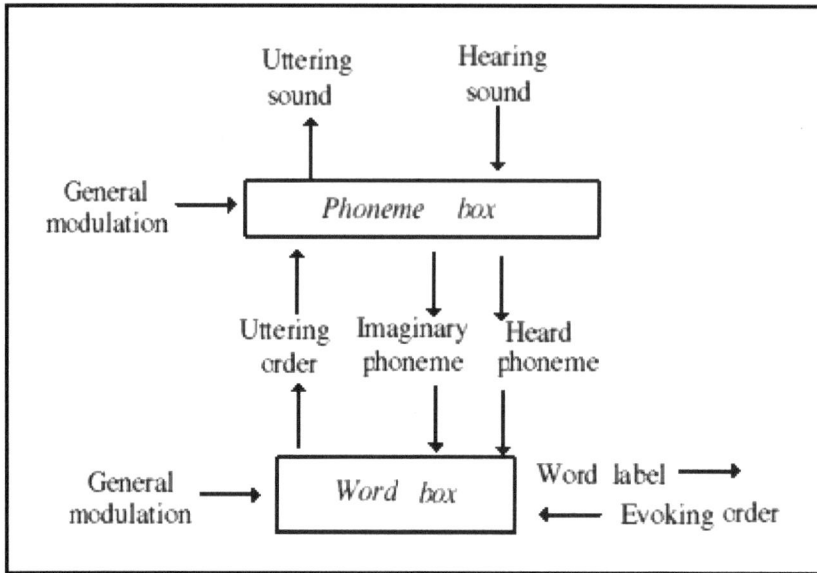

Fig. (5.3). From phonemes to words.

The heard phonemes are the detailed inputs of the word device, the orders to utter phonemes its detailed outputs. The general inputs are specific to each word: a specific order, which occurs when this word is chosen and rules its evocation, and a general emotional modulation, which rules the learning. The part of the general output (the word label) will be discussed hereafter.

Remark: The phonemes recognizing device has a weak storing capacity while the capacity of the words recognizing device is much larger. As a result, when adults learn a new foreign language, their phonemes library is saturated, while their words library is not: they will speak new words with the phonemes of their previous language.

5.1.3.2.2. The Meaning of Words

Each word is learned. The learning process was described in part IV: learning occurs when a general exciting modulation, the hearing of a word and a "meaningful signal" occur simultaneously. (Robinson shows his chest and says "me, Robinson"; Friday learns then the name "Robinson").

Language is a pyramidal sequence of abstractions arising from words acting (as in animals) as drives of some behavioral response and in the same time as detailed lines of some abstractions.

Basic Words: We distinguish very easily a human face from others and we give it a name. But it is very difficult to describe the specificity of this face by using words. This is why descriptions of faces in novels are always tedious. More generally, dictionaries cannot describe in a simple way the meaning of first learned words. (In the same way, Plato, trying to precisely define the man, proposed: an animal with two legs and bare skin; Diogenes brought to him a plucked chicken). From this, we conclude that basic words have an emotional, behavioral or sensorial meaning.

Derived Words: A derived word is the general line of an abstraction making device. Its meaning is defined in dictionaries by the set of detailed lines of this device. Other derived words can be obtained by combining them through AND, OR, INVERSE operations. Most of mathematical concepts are built in this way.

5.1.4. Words, Drives, Sentences

Main results: Speaker's instinctive drives.

Words are learned once for all and our learning capacity is limited. The number of sentences we pronounce all along our life is almost infinite; the brain cannot durably store them. With the rare exception of learned by heart texts, they are quickly forgotten. So, we have to solve two problems: a) how are sequences and long speeches generated? b) How are long texts understood?

5.1.4.1. Choosing a Word

5.1.4.1.1. The Choosing Box

Human beings only hear or utter one word at a time. This property requires a choice. As described in part III, several competitive drives excite a word-choosing box. The more intensive drive is chosen. It induces the firing of its associated output. In a second choosing box, it links with a still available column of the word device; then this column is linked with a heard word. Lastly, a direct pathway is opened from the output line to the word. Consequently, the choice of the drive triggers an all or none evoking order transmitted to the word device. On the contrary, when the word is excited, its label sends back a signal toward the drive device.

Primitive words are probably of fewer than 10^3. The same number of drives is associated with them. The other drives are associated with derived words.

5.1.4.2. The Speaker's Drives

Each competitive drive is the sum of several partial drives: one of them is the meaningful signal; we call the others secondary drives. The intensity of the total signal is the weighted sum of all the partial intensities. Some of them give negative contribution: they are inhibiting. What is the nature of the secondary drives? How are they built up?

5.1.4.2.1. A First Example: Learning a Foreign Language

When an English-speaking person starts learning German, they are said that the words "woman" and "frau" have the same meaning. In a first step, the word "frau" is linked to "woman", "woman" itself being linked to its meaning. In a second step, "frau" is directly associated with the meaning and all the drives exciting the word "woman". Soon, differentiation separates the two words. The same process occurs with the verbs "to eat" and "essen". An associative device links them, the label of which characterizes the German language. And we know that the forward excitation of some German words (or the view of the German teacher) excites backward all the German words (and favors the choice of them). This is a typical secondary drive.

5.1.4.2.2. The Secondary Drives System

It is is progressively built up from the primitive meaningful signals (Fig. **5.4**).

Secondary drives include a symmetric forward-backward circuitry. Most of secondary drives are linked to several words (*e.g.* the German drive is linked to all the German words) and most of words are linked to several secondary drives (*e.g.,* the verb "essen" to the German language and to the hunger behavior). The drivers and the words make up a multiple push-pull system (Fig. **5.4**): hearing or thinking of a word (firing of the word label) induces a back signal in several secondary devices. Some of them, which were mute, begin to fire. Their signal reaches several words, the set of total drives changes, another word is chosen. Consequently, the total drives result from a fluctuating mosaic of partial drives, a word calls for another word.

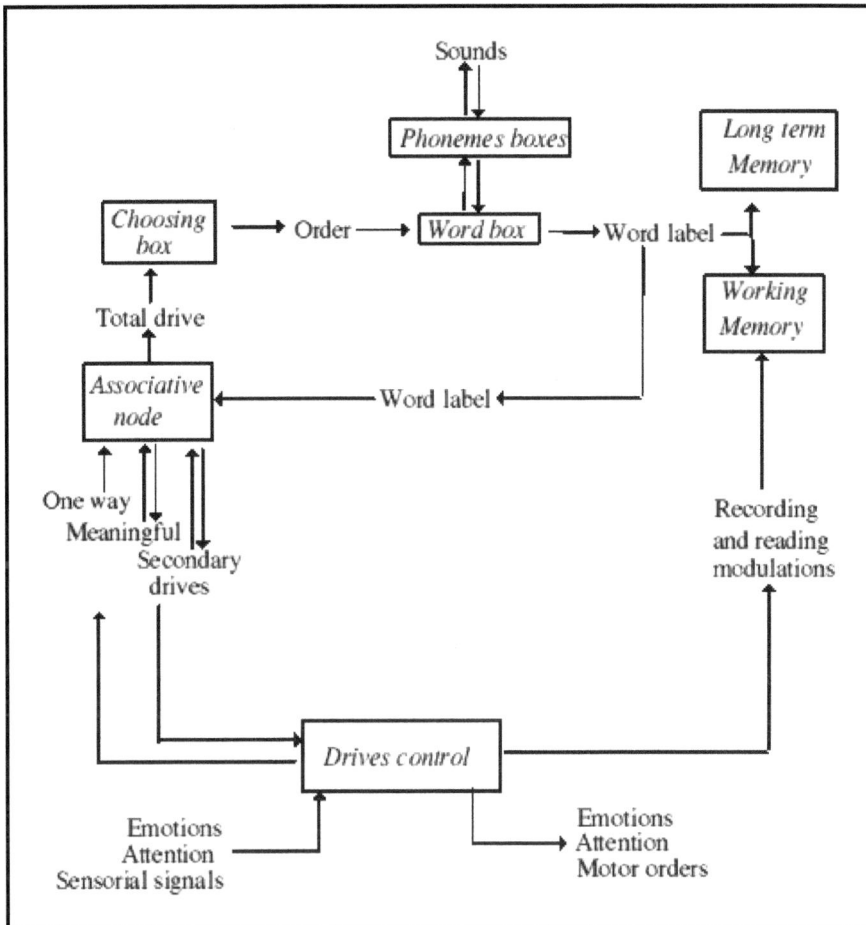

Fig. (5.4). The secondary drives help to the choice of words.

5.1.4.2.3. Control of the Drives

Words are connected through phonemes to sounds hearing and uttering out. All the other motor orders and sensorial signals are connected to the instinctive drives system and the various drives are mutually connected. (Excitation of some emotional pathways by the backward drives can trigger visual and aural evoking of memorized scenes). Once triggered, some of the drives maintain a constant intensity. The intensity of some others is slowly increasing (generalized impatience). Like all the instinctive drives, their firing is stopped by an end signal, often due to a short and strong pleasure impulse. The pleasure felt when an expected word is heard (thus ending impatience) is clearly observed in children listening to a well-known story. In the same way, the pleasure to read a thriller or a love story can be related to the excitation of an emotional process: the beginning of the story creates impatience; the ending generates a strong satisfactory signal, therefore a pleasure.

5.1.4.2.4. Associated Memories

Sequences of words can be kept in two kinds of memories:

A Small-Capacity Working Memory: In the morning, I remember the sentences I have built up during the night. In the evening, I have forgotten them. These observations are specific features of a working memory. The general modulation, which allowed recording during the night, was felt as an effort for attention. The memory clearly owns a pyramidal structure, each recorded sentence calling for the following one. This memory allows looking at the long-range coherence of the text.

A Small Capacity Long-Term Memory: It makes us able to learn some texts by heart.

5.1.4.2.5. Where in the Brain?

While the words to phonemes relation seems to be supported by the Broca's area (close to the sound uttering area), the understanding of words (the secondary drives circuitry) seems to be supported by the Wernicke's area (close to the primary aural area). Many words are stored in various places of the temporal cortex. Many places of the brain contribute to the various secondary drives. The frontal cortex provides some general modulations; frontal lesions lead to repetitions of the same sentence without progress of the thought. The limbic system provides emotional excitations. The deeper layer of the brain governs attention and vigilance, which are excited by the sensorial signals, feedbacks from the higher brain layers and emotive signals.

5.1.4.2.6. Aphasia

Forgetting Surnames: Many old people have difficulty in remembering surnames of seldom evoked personnages. These surnames are words supported by weak n_R

links. Recollection uses slowly increasing impatience. We assume that ageing involves a decrease of the maximum firing rate of this impatience, which is then sufficient to trigger the firing of recently used words, but not the firing of seldom used (weak n_R) words.

Broca's Aphasia: In many cases, aphasia occurs after a vascular cerebral accident. Broca's aphasia is a trouble of the storing words area. Hearing words (the sensorial inputs) and the understanding of words (therefore the outputs toward Wernicke's area) are normal. But recollection of words is difficult. The speech is briskly interrupted, new words cannot be found. These troubles are more severe when the patient is stressed or tired. A part of thalamus under the Broca's cortical area is in charge of the recollecting modulations. We assume that, after the vascular accident, irrigation is insufficient for hard and long working of the thalamus. When wastes accumulate, a security device inhibits the modulating center.

Wernicke's Aphasia: Troubles induced by a vascular accident under the Wernicke's area are explained in a similar way. In many cases, they seem preferentially inhibit the competitive inhibitions circuitry. Several words or phonemes are excited simultaneously, the speech is meaningless.

Conduction Aphasia: The patient cannot repeat sentences. It is a trouble of the short-term memories.

5.1.4.3. Some Examples of Drives

5.1.4.3.1. One-Way Signals

General Learning Modulations: These emotional signals enable the learning of words, the linking of secondary drives with the meaningful one, the acceptance of sequences of words by the working memory and by the long-term memory.

General Modulating Attention: Vigilance and attention enable the working without learning of the system. We will see in 5.2 the major part that the focused attention plays in creative thinking.

5.1.4.3.2. Switching Between Listening and Speaking

A general hearing modulation is excited by real oral signals. A general speaking modulation is excited by some emotions and mostly self-excitement. Both are competitive. When the speaking is switched on, real sensorial input signals are inhibited. If listening is chosen, all motor orders and all imaginary outputs are inhibited. Regarding the choice between listening and speaking, attention favors listening behavior while speaking impatience favors speaking behavior.

5.1.4.3.3. Groups of Words

Natural Groups: The meaningful signals associated with primitive words belonging to some groups (identity of a human face, mood of this face…). Each group is processed in a different place of the brain. The all or none witness of the activity of a group is a secondary drive, which favors (*i.e.,* which call for) any word of the same group. As a result, the noun "bird" favors the verbs to sing, to chirp, and to cheep and concurrently inhibits to bray. Hearing such sentences as "the blackbird brays" triggers an alert signal.

The Grammatical Testing Device: In the same way, subjects, verbs and complements belong to different groups. The first words of a child are linked with people (their mother, themselves personally), with motor behaviors (to eat, to stroll…) or to objects (the doll, the car). In a rough approximation, the first ones are mainly used as subjects, the seconds as verbs, the others as complements. Now, it was recognized that the simplest sentences in all languages are made of a subject, a verb and a complement: whatever this language is, sentences of children are ruled by a "universal syntax" [5]. Differentiation and association lead to the rule: verb after subject. From then, the firing of any subject triggers a drive exciting all the verbs (and stops the drive exciting all subjects). Thus, in French or English languages, a subject calls automatically for a verb. In the German language, the same associative device leads to the rule: verb at the end of the sentence.

Building Sentences: To summarize: the uttering of a name triggers a drive exciting the group of verbs. The drive intensity increases (an impatience). The uttering of a verb (the firing of its label) is the end signal of the drive. Several drives (grammatical rules, musical rhythm, goal of the sentences…) are combined to select a precise word, then a similar process selects the following word, and so on.

Understanding Sentences: Sentences are controlled sequences of words and speeches are controlled sequences of sentences. Thanks to introspection, we know that sentences and speeches, before being immediately forgotten, are understood as a whole. But how does this occur?

Most of them are not memorized, but spread along time by slow synapses. The uttering of a word is of about 0.3 second long. But backward drives are excited by slow synapses: they fire during about 4 seconds (the present time duration). Thus, emotional and behavioral outputs of the drive control device (see Fig. **5.4**) are not excited by one word, but by all the sequence of words uttered during some seconds. (Note that, by using a feedback not represented on Fig. (**5.4**), the firing of the word label inhibits the command for this word during about 0.5 second. Thus, the word cannot be immediately repeated. Generally, the forward drive for a word decreases (and the competitive ones increase) before 0.5 s. But in some cases, for instance when we are drowsy, the drives do not change. Then, the loop {label, backward then forward drive, word order} induces repetitions).

5.1.4.5. An Example: Learning and Reciting a Lesson

5.1.4.5.1. Listening to Short or Long Sentences

Short sentences generate immediate answers (when M declares to W: "I love you", W flushes). On the contrary, long sequences and sequences of sequences (stories) rarely generate an immediate answer. They are rarely kept in a memorizing device. Yet, the information is not totally lost.

5.1.4.5.2. Understanding the Teacher

If hearing (or reading) the teacher explanations triggers a sufficient attention, two different phenomena can be observed: a) learning by heart: the pupil will recite his lesson without changing a word and b) activating a new associative circuitry. For instance, the name Hitler becomes linked through some secondary drives to the words war, Jews, Stalingrad, Normandy. From then, hearing the name Hitler causes the triggering of the associated secondary drives. The pupil will make original sentences to utter the learned names: we say he has understood the lesson.

5.1.4.5.3. Uttering Original Sentences

This is an example of sentences making. The main drives are the need to speak, a focused attention, and the usual grammatical drives. The drives associated to the words war, Jews… are firing (and their intensity perhaps increasing). For each of them, the satisfactory event triggering the end signal is the hearing of the word. But grammatical rules forbid to call one after the other, they have to be separated by correct sentences.

5.1.4.6. The Device at Work

5.1.4.6.1. To Hear or to Speak

The need to speak is obvious in many adults. In these people, a strong impatience excites the general speaking modulation; they cut short any other speaker. Uttered words stop the repeating device; thus, these people only have short thoughts. Their chattering is a continuous flow. On the opposite, the utilitarian language is favored by general attention and by a strong hearing modulation. A pilot and his navigator only exchange some simple words (right, left, slowly). In case of failing attention, inner speech works in an autonomous way.

5.1.4.6.2. Some Examples

Wandering Inner Speech: Without any external drive, each word calls for the following one without any constraint.

Reading a Thrilling Novel: The visual attention favors Foveal vision; lateral vision and hearing are inhibited. The inner speech extracts abstract messages from the sequences of recognized words. These messages excite imaginary emotions, inner speech attention and backward sensorial attention.

Looking at a Tedious Movie: The sensorial attention is not sustained. From then, the inner speech works as an autonomous loop.

Compelled Attention: A teacher forces us to read Spinoza. The order uttered by the teacher triggers our sensorial attention. We begin to read. The word system does not extract any emotional message. Inner speech attention is not sustained, neither the visual one. The colliculus makes another choice: we look at the flies.

5.2. Action, Thought and Self Awareness

Introduction: Everybody feels that thought and consciousness exist, that they are the most precious part of our nervous activity. But nobody knows exactly what they are. Consciousness is sometimes defined as the feeling, the more or less sharp intuition of our state and of our acts, as an immediate knowledge of the realty and of the external world. Thinking has a more rational appearance. It is responsible for rational reasoning, choice of a strategy, imagination and creativity. But many thoughts are obviously conscious; consciousness and thought are intimately related. Both depends on our emotions, on the "me personally" signals, on most of the received sensorial messages. Both are more or less associated with memory and with language. Both are linked with past experiments, with the ability to foresee the consequences of our acts, to make "mental experiments". Note lastly that in most of languages, the moral sense and the human self-consciousness are pointed out by the same word.

The pioneer work of Penfield [6], who systematically studied the mental effects of war wounds, enlightened for the first time the relation between neurophysiology and thought. Today, 3D imagery allows locating small cortical lesions while sophisticated tests detect the associated mental defects. For instance [7], lesions of the orbito-frontal cortex are associated with a loss of self-control. On these bases, confirmed by some studies of "non-human primates", we assume that the words "consciousness" and "thought" are related to a set of precisely defined functions which can be localized, most of them in the frontal cortex. We have now to show that these functions are supported by neuronal circuitries similar to those we have studied above.

5.2.1. Complex Motions

Main result: Ghost actions and language are supported by similar structures.

5.2.1.1. Some Examples of Increasing Complexity

5.2.1.1.1. Simple and Complex Actions

Men not only speak; they act. The combination of imaginary speeches and acts is close to what we call the thought. We distinguish simple actions (for instance taking soup in a spoon and lifting it to your mouth) from complex actions (drive to the store, buy a can of soup, come back, open the can and put it on an oven…). The examples of the next paragraph will make the difference clearer. In short, simple actions or sets of several simple actions lead to stereotyped behaviors. Many animals, if not all animals, carry out simple actions (ants bring back some pieces of leaves, birds build their nest). Complex actions lead to innovating behaviors. Only men are capable of very complex actions.

5.2.1.1.2. The Frightened Squirrel

An emotional drive (the fear) is chosen; it triggers an action order which governs a sequence of three inborn motions: to run, jump and climb. Each motion is self-adaptable: virtual motions enable first trial and error correcting before choosing the direction of the targeted tree. Then, the real motion is carried on. Its end triggers the following motion. The result of the whole action (the success of the flight) stops the drives.

5.2.1.1.3. My Daily Journey

The number of inborn motions (like the number of recognized phonemes) is fairly small. Men are capable of learning some new motions such as riding a bicycle or zapping the TV. We easily conceive pyramidal stacks of elementary motions. Thus, each morning, I take a bus, then the subway to my office. This learned complex sequence of simple actions, which probably has a pyramidal structure, has become an automatic routine.

5.2.1.1.4. The Spontaneously Performing Dog

Learned pyramidal structures do not explain the diversity and the properties of the observed complex actions. We have seen in 5.1 how a poodle learns how to sit up and begs when it is near its owner and hears a verbal order (the action drive is an association of the owner's smell and a specific aural signal). The result of the action (its success) is linked to a reward (a stroke, applauses) which triggers an emotional signal. Sometimes, the dog wishes to be stroked (an emotional signal is sent backward); it spontaneously sits up and beg: the virtual reward sends back a drive signal and the associative adding of this signal to the master vicinity is sufficient to trigger the action without verbal order. The main new fact is the use of a virtual result as a component of the action drive.

5.2.1.1.5. Doing Some Shopping

Needing some food, I decided to go to my parking lot, take my car and drive to some store: a very complex sequence of simple motions. To quickly conceive the whole sequence, I did not use imaginative trials of the motions. I just linked the order (and the label) of each simple action to its expected result (such as the opening of the car doors) without looking at the way to obtain this result. The main new facts are the use of intermediate results (without any direct emotional link) and the bypassing of real and imaginary motions.

5.2.1.2. A Convenient Architecture

There are striking similarities between complex actions and language. We search therefore for a similar architecture. Fig. (**5.5**) is a transposition of Fig. (**5.4**).

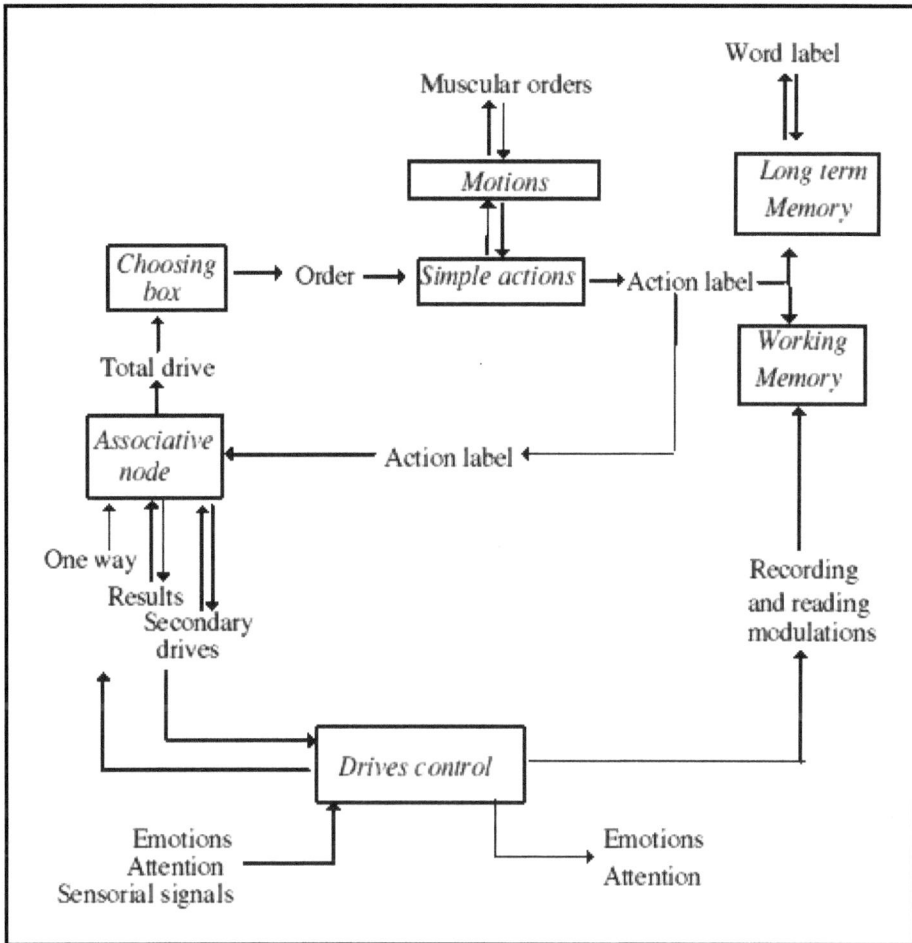

Fig. (5.5). The control of complex motions.

In this scheme, motions replace phonemes, simple actions replace words and complex actions replace sentences. The action label (as the word label) is supported by the general line of an abstraction making device. The architecture exhibits a systematic symmetry between forward and backward pathways.

5.2.1.2.1. Ghost Motions

The simple action and the motion boxes have obviously two working modes: 1) real motions (the muscles are excited) and 2) imaginary motions (while trial and error correcting preparation of real motions). A third operating mode appears when, after learning, a bypass links the order to the label. The action and the motion boxes do not intervene. We call this mode a "ghost motion". It instantaneously works while the implementation of the real or the imaginary modes requires some time.

5.2.1.2.2. Result Drives

End signals (results) of simple actions replace first generation meaningful drives. Thus, performances of simple actions and their results are linked by forward and backward direct pathways.

5.2.1.2.3. Paradoxical Inhibitions

Some inhibiting drives prohibit the choice of all the actions incompatible with the present situation (for instance walking ahead when I am on the edge of a cliff). We will see (in 5.2.5) that they do not work like the normal (first order) inhibitions.

5.2.1.3. The Device at Work

5.2.1.3.1. A Simplistic Case

Someone knocks at the door. The emotional signal triggers a wished result (a goal): to approach the door. This wished result acts as a forward drive, the walking action is chosen, the label of the ghost action acts as a backward drive meaning that this action would lead to the required result. The drives control orders to pass on the real mode.

5.2.1.3.2. A More Complex Case

Someone knocks at my door when I am on a ladder. The group inhibition forbids walking directly to the door. Therefore, there is no simple possible action linking the present situation to the wished result. Increasing impatience excites adjacent drives. Thus, the device tests sequences of two simple actions. Repeated experiments lead to associative linking with secondary drives. If, in a later experiment, a two-step sequence cannot be found, the device tests three-step sequences. The secondary drives tend to favor the useful two-step sequence. Thus, in the first time, the device searches for a third sequence added to a previously used two-step sequence; this method reduces the number of sequences to be tested. Note that the device does not select the best sequence but the first that has been found, showing that we have many non-optimized habits.

The sequence of signals, similar to an inner speech, is exploited as a whole (as a sentence) by a trial and error correcting process until a convenient sequence of simple actions is obtained.

5.2.1.3.3. Working Memory

When a ghost sequence is accepted as convenient, the real working mode is selected by the drives control. But, as for verbal thought, the time scale of devising an action is of about one second while the time scale of carrying out a real action is often of an hour. The planning of a complex action has to be stored within a working memory during the slow accomplishment of the successive motions. If the memorizing modulations are too weak, the goal of the sequence is forgotten before the sequence is achieved. This is why, for instance, old people forget where they are going.

5.2.1.3.4. Long-Term Memory

If often repeated, some long sequences of simple actions (for instance my daily journey) are memorized. They are automatically carried out, without any thinking.

5.2.1.4. Combining Language and Action

Language is mainly supported by the left cortical hemisphere whereas action is mainly supported by the right hemisphere. Both are linked by the corpus callosum and probably by some thalamic pathways. The troubles exhibited by split-brain patients [8] clearly show that, in normal people, language and action act together.

We assume that the labels of sequences kept in the long term memory can link forward and backward with specific words (or with specific words in short sentences: the word "car" in the sentence "driving my car"). Then, the order, the report or the evocation of very complex actions can be processed in a marvelously compact way: "I am going to my office" or "I am building a house" sums up huge and intricate sequences of complex actions. On the other hand, complex actions are related to a goal (a wished result). Thus, words can be associated with goals: evoking a word can become a goal; emotional excitation of a goal can excite a word label. Several words can be linked by associative devices; they can give rise to differentiation; they can form time-delayed sequences.

5.2.1.5. Verbal Action; Long Term Plans

The expression "verbal action" does not describe motions of the speaking process, but any verbal description replacing the sensorial report of my real actual position. We just demonstrated that a verbal goal can take the place of a wished result. So, an acting process (working only in the ghost mode) can lead from a verbal situation to a verbal result. Such verbal actions support, for instance, long-term plans. Note that real actions act within present time while the combined language-action devices, although very fast, focus on an indefinite future.

5.2.2. Consciousness

Main results: Conditions for consciousness. Role of focused attention.

5.2.2.1. A First Attempt to Define Sensorial Consciousness

Consciousness is an obvious introspective fact. But can we specify its neuronal support? Discussions between materialists and "dualists" are still virulent. Necessity to start from experimental facts is now well understood [9]. But, for only this particular sublect, objective methods (observation of monkeys, functional imagery of the brain, subtle tests) are not convincing. We do not escape to the discussion of introspective reports. We clearly distinguish sensorial consciousness from self-consciousness (I am myself) and from moral consciousness (the sense of good and evil). All the sensorial signals reaching the cortex are not conscious. I read a fascinating novel; my reading is conscious. I read a bothering one; I cannot remember a word of it. Other example: Going out, I switch off the light. But on the stairs, I wonder if I did it. Thus, my gesture was unconscious.

From these examples, we infer that consciousness is not associated to the gesture but to the remembrance of the gesture. In the case of the novel, it is not associated to the reading behavior, but to the understanding of the text: a signal is conscious if it is commented and linked with other conscious feelings.

5.2.2.1.1. Always Unconscious, Possibly Conscious and Actually Conscious Signals

As a first attempt to define sensorial consciousness, we assume that it is strongly related with language device (and, by symmetry, with the complex actions device). The word devices are not excited by primary sensorial signals, but only by phonemes. More generally, the language-action system only works with processed sensorial signals (with recognized objects, with the moods of people and not with the first visual signals). Thus, all signals used at the spinal or at the cerebellum levels can never be conscious. For instance, when I burn my hand, a fast pain fibre induces a brisk motion. This pain signal and the reflex order are not conscious. But other fibres send a pain report to the brain, some sensors send a motion report, and these reports may be conscious.

Assuming that sensorial consciousness and the language device are strongly related, all inputs in the language action system (the external drives governing the system and the internal drives resulting from the inner speech) <u>may</u> be conscious. This is a constant property of these inputs. On the other hand, these drives are competitive: only one word (or one action) is chosen at a given time, and the drives associated with the chosen order can be commented and thus be <u>actually</u> conscious. The other drives are then unconscious. Choices are constantly evolving. As a result, actual consciousness is a constantly varying fluctuating property.

5.2.2.2. *Some Examples of Possibly Conscious Signals*

Animals own probably some sensorial consciousness: look at a cat computing a jump. In this case, consciousness is related to imaginary motor experiments (carried out by the premotor cortex). In human beings, consciousness seems widely related to the inner speech. More precisely, the imaginative hearing of words is conscious, not the very fast ghost mode. Dreaming and musing (which are inner speeches without strong attention) are probably not fully conscious.

As expected, many possibly conscious signals come from the cortical processing areas (for instance words, identity of recognized people, mood of the people, nature of recognized objects, body posture, and bladder volume). But other possibly conscious signals originate from deeper brain strata (for instance pain, fear and other emotions). And when Marcel Proust remembers Combray (see 4.2), he described his painful attention effort, hence a conscious feeling.

Instinctive drives are possibly conscious, for instance the drive of thirst (see 3.2). The consciously felt signal is not the osmotic pressure of blood, but the total drive. Note that a component of this drive is subjective, therefore generated by the language and complex action devices: the thirst feeling increases when we see a fountain.

These assumptions are supported by the results of experimental cognitive psychology. Human consciousness is related to a "global neuronal workspace" [10] which is mainly a large thalamocortical complex [11]. Experiments show [12] that, unlike unconscious stimuli, conscious stimuli create a long excitation in a wide part of the brain.

5.2.2.3. A Disturbing Observation

In our first attempt, we assumed that a signal (the drinking need) becomes conscious when the drinking behavior is chosen by the complex actions device. This assumption must be corrected: in the desert (where drinking behavior cannot be chosen), thirst is strongly felt.

5.2.2.3.1. Control of Bladder Voiding

The same phenomenon occurs in another simple example of conscious/unconscious interacting: the voluntary control of bladder voiding. The bladder volume is continuously increased by the production of the kidney. The bladder muscle and the striated sphincter are driven by a two-position switch (located in the pons): in the waiting position, the sphincter is excited and the muscle is not excited; in the voiding position, the muscle is excited and the sphincter is not excited. This is an unconscious device. At rest, a spontaneous neuron keeps the waiting behavior on. An excitation originating from the conscious brain is required to switch off the waiting order and switch on the voiding order.

Sensors included in the bladder generate a possibly conscious drive increasing with the bladder volume. I am watching a thrilling movie; the bladder signal is not the most intensive drive, it is not chosen: I do not feel any voiding need. The movie ends and in the simplest circumstances, the bladder signal is chosen; I begin to feel a voiding need. Things are often more intricate: voiding here (in the theater) is forbidden by an inhibiting drive, the voiding behavior is not chosen; yet, voiding is kept as a goal, an increasing impatience is triggered. This is felt as a conscious and soon imperious voiding need till I find the toilets.

5.2.2.4. Paradoxical Inhibitions, Focused Attention

This phenomenon requires splitting the inhibiting drives into two groups. Common drives compete in the words (or the simple actions) choosing box. Paradoxical inhibitions are associated with forbidden or impossible actions. They are connected to the circuitry between the choosing box and the word box (Fig. **5.6**).

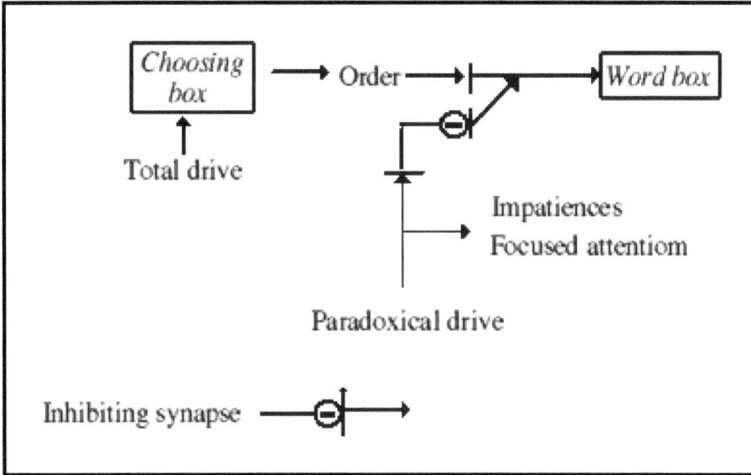

Fig. (5.6). The effects of a paradoxical drive.

If the word is chosen and if a paradoxical inhibition fires, then the following steps are taken: a) The word is inhibited AND the imaginary or real modes of other words are strongly inhibited; b) Strong impatience excites the ghost mode of the associated words and inhibits the others, leading to an exasperated and focused search for allowed speech.

These effects appear as a focused attention, which can touch the response to any main drive from a deep behavioral drive (sex, food) to the more abstract concept (a mathematical conjecture). When their attention is focused, people think strongly and exclusively about sex, food or the demonstration of their conjecture.

5.2.2.5. Final Assumptions

To summarize this discussion, we infer that any processed sensorial signal becomes conscious when it is associated with a strong attention. Only the imaginary and the real modes are conscious, not the ghost mode. Thus, we "hear" the words of the inner speech. This property requires that the signals are stored in a working memory: real or imaginary, conscious signals may then be commented and linked with other conscious feelings. The strong attention is probably used as a modulating excitation to store the conscious signals

5.2.2.6. Self Awareness

It is supported by the message "Me personally" (and by the inverted message "Somebody or something else"). A neuron firing generates a message meaning «such a muscle or such a limb is voluntarily moving» if the sensors detect a motion AND if the muscle is excited. The association of several such signals leads to the abstraction "Me personally". Some other sensations contribute to this signal. For instance, the strong intensity of an attention (strong intensity due to impatience or to an obsessive state) leads to the distension of the local blood vessels, the sensors of which generate a feeling of painful effort. The result is the message "My own painful thinking".

5.2.2.6.1. Self-Consciousness is Useful

Predictive experiments result from a continuous mix-up between my own emotions and the emotions of other people, between real and imaginary motions and sensations. A terminal filtering seems necessary to avoid errors: operations AND between the main outputs and the signal "Me personally" allow avoiding delirium.

5.2.3. Fiction, Creativity and Rational Thinking

Main results: Our behaviors are governed by fictional goals and opinions. Creativity results often from the decrease of an obsessive drive.

5.2.3.1. Speech and Fiction

We just demonstrated the part of paradoxical drives in some actions. We will now study their role in complex thought. Some of the paradoxical drives are learned (do not tell cuss words before a lady). We have seen that other paradoxical drives are generated when the devices do not find a convenient ghost action (for instance, when we do not find the word expressing our thought).

We will see in the next Section how our long term goals and our opinions govern our behaviors. But we will first show that our behaviors govern our goals and beliefs.

We have seen that, if a one step action is found impossible, the device does not stop, but paradoxical drives trigger increasing impatience till the finding of a more complex convenient action. This property seems fairly general: the verbal and the action devices overcome difficulties by discovering new, more complex and often fictional solutions. The observation of neurological patients supplies some striking examples [13]. Many other examples of this important phenomenon are found in normal people:

5.2.3.1.1. Neologisms

During real speech, a word cannot be found; an impatience leads to the invention of a neologism.

5.2.3.1.2. Fictional Goals

Some behaviors are due to irrational chronometric drives (for instance, the need to run) or sometimes to physiological needs driven back into the freudian unconscious. In both cases, the device cannot work by lacking a goal. It triggers impatience increasing till a fictitious rational goal is found.

5.2.3.1.3. Fictional Word Meaning

Some words are not supported by any sensorial meaning, for instance the word Xanadu. The device supplies the fictional country referred to by this name with dreamed descriptions.

5.2.3.1.4. Conflicts Between Fiction and Reality

I travel within Xanadu after having dreamed of it. In many cases, the sensorial analysis uses discrimination, real and dreamed signals to go up through two different pathways. But sometimes, reality is so frightful that its pathway is inhibited. Then, how rational may the arguments be, one believes the fiction and not the reality.

Thus, any signal linked to a strong impulse (for instance, to obtain a tax exoneration) becomes a possible goal, which induces a rational justification usable as a moral rule.

5.2.3.2. The Inventive Process

Search for proper nouns, for complex actions and invention results from the associative performances of successive periods of strong and weak focused attention due to paradoxical inhibitions, while focused attention excites a part of the speech system and inhibits another part.

5.2.3.2.1. Searching for Proper Nouns

Some people (and especially old ones) cannot recall proper nouns. These nouns have to be extracted from some abstract source ("this Martin something, remember, the druggist's nephew...") but the pyramidal process does not find any solution. The growing of generalized impatience is clearly felt. In this case, the mental experiment is a failure because the searched noun is in the part of the system inhibited by the strong focused attention. After disappointment, its intensity decreases; the inhibiting effect disappears while some traces of the exciting effect remain for a few minutes. The searched noun is remembered during this phase.

5.2.3.2.2. Short-Term and Long-Term Focused Attention

Usual impatience's (leading for instance to neologisms) are signals growing during some seconds. If they are, as in the previous example, relayed by some periodic neuron, they can govern attention for a few minutes. Invention, a rare and slow process described by the great mathematician Henri Poincaré, is obviously related to an effort lasting much longer. We assume that long-lasting attention is due to long term potentiation (see 4.1.2) with a time constant about one day. Thus, we need one day to focus our attention on a particular subject and one day to get free of it.

5.2.3.2.3. Sequences of Strong and Weak Focused Attention Intensity

When our attention is focused, it strongly inhibits a great part of the speech system and this inhibition forbids unexpected associations. During drowsiness, the attention decreases. Its inhibiting effect completely disappears while some traces of the exciting effect remain. Then, unexpected associations may occur and be agreed by an associated trial and error correcting device. This is why Newton's apple has been associated with an action without reaction; mental experiment identified at once the lacking centrifugal force as the lacking reaction in the rectilinear motion of the fruit and, in a flash, this idea was recognized as the solution.

5.2.3.3. Inventing and Using Tools

The invention of a new machine (for instance the reputed impossible electric motor using continuous current, the Gramme's dynamo) results from the same psychological crisis as the mathematical invention. But, while the mathematical thought is mostly abstract, the invention of a machine requires both theoretical thinking (due to the verbal system) and practical thinking (due to the complex actions system).

Making use of a tool is a learned motion. Use of fork and knife is integrated in a simple action (to eat). A chimpanzee could probably be trained to use them. On the opposite, the use of hammer and graver by a new Michael Angelo which requires visual imagination is supported by the complex actions device. First, in ghost mode, the goal of the hammer stroke is deduced from the global goal (the wished statue) and from the visual report of the actual appearance of the piece of marble. Then, in imaginative mode (which allows trial and error correcting processing), the simple action device searches for the convenient adjusting of the elementary motions. Lastly, real mode is switched on, the hammer hits the graver and the processing begins again to control the next hammer stroke.

The use of a machine (my car) is often supported by both the action and the verbal devices.

5.2.3.4. Abstract Reasoning, Deduction and Induction

5.2.3.4.1. Deductive Processing

Abstract reasoning appears as a variant of speech. It is supported by most of the neuronal functions above-mentioned. Suppose that we have to prove a mathematical proposition. Building a correct thinking process to achieve this goal is similar to building a complex motion through mental search or enunciating a correct sentence in a language. Starting from an available knowledge, we have to find a way that reaches our goal in accordance with some rules that have been acquired either by teaching or by experience. As in the case of usual language, through the use of mental experiments and associative links, several "ghost motions" can be devised. A sequence of ghost messages exciting a sequential box corresponds to a chain of intermediate conclusions in the thinking process. Mental experiments excite sequences of abstract behaviors or assertions that try to link the expected result to the available knowledge. Unproductive experiments are eliminated by a verification process. Partial verification or proof of each assertion takes into account the set of learned rules. The tree-like set of proven assertions admitting the ultimate goal as its root is a correct reasoning process. The logical or deductive part of the abstract reasoning process is equivalent to the syntax of a language.

5.2.3.4.2. Inductive Processing

But if deduction is good at asserting the validity of the conclusion of a thinking process, it is of limited help in finding new ideas or developing new concepts. For instance, the negation of the axiom of the parallels in the Euclidian geometry, a first logical step, does not imply the assertion of another axiom for replacement: Riemann or Lobatchevsky had to imagine and define their own axioms to build their new geometries. Inductive process is a variant of creative process. Mental experiments have to be designed and conducted before being verified. But imagination has to be excited and guided by the recognition of some similarities in the enunciation of a new problem with other problems that have already been solved. Variations around the reasoning processes used to solve those problems will be used to build the mental experiments. These variations are often obtained by reaching a higher level of abstraction that is another floor of the neuronal pyramidal structure supporting speech capabilities.

Three steps are needed to reach a new result.

a) Inventing new goals (To classify plants …) and new methods (To define a tree approach for the classification…). This step is typically a creative one. This is why it is very rare: more than 2,000 years separate the first study of series (sum of an infinite number of terms) by Zenon from the conclusion by Cauchy.

b) Conducting real or mental experiments on some examples for enlightening some property, which is guessed as a general one: the three bisecting lines of a triangle have a common point; the sum of the 5 first terms of series gives a good approximation of the whole sum. This intuitive guessing of a general property is called a conjecture. From the previous Sections, we know everything about the real and mental experiments. We just need to develop how neuronal circuitries isolate an apparently general property. (We are to show they use causal links).

c) The last step is the abstract reasoning conducting to a proof. We have just shown above that this last process has the same structure as speech.

5.2.3.5. Links Between Cause and Effect

The "Mc personally" signal has already appeared in the description of how the posture of the body is controlled. It is the output of the associative linking (operation OR) of all the outputs from differentiating devices linking some motor order AND some change of the body posture, taking into account that associative devices have backward pathways. Thus, unexecuted motor orders and simultaneous firing of "Me personally" induce the imaginary feeling of a change of posture.

Likewise, the output of the associative linking of all the outputs from differentiating devices linking some motor order AND some change of the external world is a "cause and effect" special signal of which the meaning is "therefore". As above, backward pathways make that an unexecuted motor order and the simultaneous firing of the "therefore" signal (the equivalent of "Me personally") generate an excitation of the effect neuron (or of several competitive effect neurons; only one of them will be chosen).

When we draw straight lines from the summits of a triangle, we obtain one central triangle with no side included in a side of the main one. It appears that this triangle may be reduced to a point when the straight lines are the bisecting lines. The primary visual analyzer distinguishes a point from a triangle. Then, a cause and effect link associating bisecting lines to a "therefore" assertion takes place. The three bisecting lines have a common point. At this stage, mental experiments are carried out to try to link the knowledge we already have of the properties of the triangle and its bisecting lines to the "therefore" assertion (induction or intuition). Abstractions built for such links are new elementary deduction. An experimented link of abstractions that never contradicts the deduction rules is selected as a valid abstract reasoning in the same way as a sentence or a path to a destination is accepted.

Note that rational reasoning is built not only from the results of real experiments, but also from the results of abstract ones.

5.2.4. Self-Control and Sociability

Main results: The frontal cortex confronts immediately most pleasant and long term best behaviors

5.2.4.1. Short-Term and Long-Term Optimizations

How does thought change our behaviors? The words "good consciousness" and "moral sense" are almost synonymous. What is moral sense?

I am longing for chocolates. But I have diabetes; the doctor advised me against eating chocolate. My self-control enables me, sometimes, to push back the temptation. More generally, the frontal cortex is in charge of a confrontation between short-term and long-term optimizations of our actions. Most of our decisions (to switch on the lamp now or later) are of no importance. Then, the simplest optimization (the short-term one) is the best. But in some cases (I know that a sniper is waiting for me), the long-term optimization has to be chosen.

5.2.4.1.1. The Immediately Most Pleasant Behavior

While mixed language-action devices consider an indefinite future, pure action devices act in the present time. Most of their action drives have an emotional origin or are closely connected to the pleasure impulses. As a result, the short-term optimization aims at the satisfaction of the more intensive drives, such as anger, need to speak, sex or at obtaining a pleasure impulse. Babies, who do not have learned goals, search for the immediate satisfaction of their basic instincts.

5.2.4.1.2. The Components of Long-Term Optimizations

Inborn Instinctive Inhibitions: For instance, look at the instinctive restriction on murder in many carnivorous mammals such as dogs or men. But in some of them, such as male lions or bears, the same restriction does not apply and these animals may kill and eat their kids.

Learned inhibitions (medical diet, moral rules)

Excitations and inhibitions due to focused attention

Excitations and inhibitions due to long-term planning

This component obviously depends on the predictive ability of the subject.

5.2.4.1.3. Ulysses and the Cyclops

When a child is asked: "What is your name", a word calls for another and the child answers: "My name is Joe Smith". This automatism results from a simple verbal automatism. But when the Cyclops asked: "What is your name", Ulysses answered after a short hesitation: "My name is Nobody": he reflected before answering. This requires a long-term optimization supported by the prediction of the Cyclops response.

5.2.4.1.4. Various Forms of Prediction

Visual Prediction: Forecasting the future position of an object from its present place and velocity is an easy extrapolation. Babies learn it very soon. But a dog, when chasing a rabbit, does not forecast the future position of its prey; the dog run toward the actual position of the rabbit. The chess player anticipating the next moves uses probably some kind of visual prediction.

Prediction and Extrapolation when Playing Tennis: When dealing with people, long term strategies must take into account the reactions of the other people and therefore you need to predict their reaction. Look at the simple example of tennis. After throwing the ball, the tennis man observes its motion. He imagines he is receiving it. Then, he imagines he throws back the ball, imagines its motion with the eyes of his partner, then with his own eyes and gets ready sometimes before the event to receive and throw back the ball. This short term mental predictive process uses continuous switching between real and virtual motor orders, real and imaginary visual inputs, "Me personally" and "Not me" signals.

Emotional Predictive Extrapolation: The same description applies to the observation of facial expression, as that facial expression displays emotions. Switching between my own emotions and the emotions of another person enables me to predict the effects of my words or my motions on this person.

5.2.4.2. Where in the Brain?

5.2.4.2.1. Evaluation of the Immediately Most Pleasant Behavior

Emotions are generated in the limbic system, the pleasant ones mainly in the septum and the unpleasant ones mainly in the amygdalae. After a relay in the medio-dorsal part of the thalamus, the signals are mostly projected on the pre-frontal cortex. The mental experiment which enables the definition of the most pleasant behavior uses only signals coming from the septum.

5.2.4.2.2. Evaluation of the Constraints Associated with Long Term Optimization

The predicted reactions of other people (probably originating from the temporal cortex) and the set of learned rules, rewards and punishments (it takes into account the signals coming from the amygdalae) contribute to the evaluation of the best long-term behavior. This operation seems to be localized in the dorso-lateral part of the pre-frontal cortex: the best long-term behavior and the immediately most pleasant one confront each other. It seems to result in the evaluation (but not to the effective decision) of the best choice.

5.2.4.2.3. Choosing the Best Behavior

Evaluating the best behavior is an operation separated from deciding what is the best behavior, which is carried on by the ventro-medial pre-frontal cortex: subjects suffering from a lesion in that area know and can say what the best decision is, but choose the wrong one [14].

The frontal cortex seems to be devoted to the detailed repercussions of choosing a behavior: the best long-term behavior inhibits efficiently most of the instinctive drives. Apparent anger and apparent greed disappear, and this is self-control. Speech is not used then for the satisfaction of the speaker, but as a tool planned to influence other people. Reminding that the long term best behavior depends on learned rules and learned punishments, we understand that what is called moral sense is the simple consequence of education and of a robust self-control.

Lesions of the frontal cortex impair the inhibition of the most pleasant behavior; they generate terrific losses of temper, unquenchable volubility and more generally cognitive impulsiveness [15], which is the impossibility to wait for the satisfaction of any caprice. Partial lesions cause partial losses of self-control, which generates stupid behaviors (for instance spending in one night all the money for the month) and are unbearable for other people (*e.g.,* awful angers). As a result, frontal impairments induce a lack of sociability.

5.2.5. Miscellaneous

Main results: a) The Freudian theory gathers real, but disparate phenomena; b) Dogs seem to own a restricted but undeniable thought.

5.2.5.1. Freudian Phenomena

While describing instincts, we recognized a century after Freud the necessity to introduce drives to explain the observed phenomena. But we noted two main differences between generalized drives and Freudian ones: a) we are dealing with a great number of "small" drives (for instance, those governing the choice of words) while Freud only looks at some "big" drives and, mostly, at libido. b) The physical nature of a Freudian drive is unknown (and called "energy" by a poetic license). On the contrary, the physical nature of a generalized drive is precisely known: its intensity is the firing rate of a peculiar neuron excited by a specialized device generating impatience and the end of drive signals.

In a general sense, Freud's work appears as a superb attempt to deduce some functional properties of the brain using only the observation of human crisis behaviors. Although the definition of functional properties faces some ambiguities and their gathering into a corpus is questionable, most of Freudian phenomena exist and are taken into account in our description.

5.2.5.1.1. Detailed Discussion

Freud distinguishes three components in the mind: the ego, the id and the super-ego.

The unconscious id contains the drives, and first of al, libido, which is sometimes defined as a "sexual energy" or sometimes as the search of any form of pleasure derived from the body. Keep in mind that, for human beings, pleasure impulse is a short signal generated by the end of drive.

The conscious ego governs the self-created interactions between the id and the super ego. The super ego is an unconscious screening mechanism which seeks to limit the blind pleasure-seeking drives of the id by the imposition of restrictive rules. (Thus, the super-ego would be the device matching the best behavior against the most pleasant one).

This matching generates serious conflicts. (For us, any choice results from a competition which, most of times, is a mild conflict). The greatest success of the Freudian theory was to highlight the defense mechanisms (repression, regression, sublimation) preventing conflicts from becoming too serious.

Repression is an inhibiting mechanism. It can act in two places: in the definition of the best behavior or as a widened inhibiting drive. In the first case, a behavior (raping or sometimes any sexual activity) is forbidden. In the second case, the conscious evocation of some subject is suppressed. (More precisely, the first conscious word induces the inhibition of a wide conscious area). But the associated instinctive drives are not eliminated. They have become subconscious.

The inhibition of a behavior reacts on the regulation of drive intensity. We have seen, when studying obsessive modulations, how some serotonin synapses increase the intensity of frequently excited drives. In the same way, the intensity of seldom excited drives decreases: fasting induces a loss of appetite. The same applies to sexual activity. This observable fact is not included in the Freudian theory. Freudian defenses appear when a behavior is frequently excited while its outputs are strongly inhibited. Now, keep in mind that a) the activation of glutamate synapses slowly decreases when they are unused; b) most of the behavior controlling devices result from several differentiations and c) each new differentiation requires the activation of links inhibiting the old output and some competitive outputs. Most of the time, these alternative pathways are not inhibited by the repressive signals. Then, the loss of activation enables the reopening of the old pathway (and this is regression) or of a competitive pathway (this is sublimation).

5.2.5.2. Are Dogs Thinking?

Between the frog (which is governed by simple automatisms and does not think) and the thinking human being, do the animals have some intermediate form of thought? We will only look into the example of dogs.

The primary sensorial signals of dogs and men are different. Sense of smelling is important in dogs, not in men. The frequency analysis of sounds in men is specialized to recognize phonemes; in dog, it is less specialized. This explains why a dog is capable of recognizing the particular noise of its owner's car at a great distance.

Dogs have all the basic instincts and especially developed social ones. As in men, their limbic system generates emotional behaviors such as rage or submissiveness.

Dogs do not stand up and do not have hands. So, their motions require a less demanding control and their cerebellum plays a less important part than in men.

In dog as in men, motions are governed by a pre-motor, a supplementary and a motor cortex. Dogs have memorizing devices. A dog is able to imitate its mother, to learn new motions and sequences of elementary motions (for instance to go and fetch the slippers of its owners).

Observing how cats and dogs choose their return pathway, we conclude that dogs and not cats are capable of predictive abstractions. When in a new place, dogs and cats explore their domain: they learn to recognize the sight or the smell of several beacons, the position of the closer ones and they associate a visual or smelling label to each of them. Once it knows the place, a cat always follows the same way from a beacon to the next one (it has learned the sequence of the labels). If I am on its road, it prefers in spite of its fear, to pass between my legs rather than losing its track. On the contrary, a dog is able to invent a new way to go home. To do so, it uses its knowledge of the beacons (a dog lost in an unknown forest cannot find its way back to its starting point). Thanks to a mental experiment, it finds the sequence of beacons defining the shortest way home. But, without language, dogs lack of available markers; its pyramid of abstractions cannot acquire new floors. It can only build up short term strategies.

As in men, the attention of dogs can disconnect a self-maintained imaginary loop from any sensorial input. Does the dog have sweet-smelling dreams?

As in men, the signal "Me personally" can be built up in dogs from associations between motion orders and sensor signals. And, as in men, the controlling motion devices introduce some mix-up between "I" and "he". Thus, the dog needs a device to separate them. From these observations, we conclude that the dog has some self-consciousness. As an opposite argument, it seems that only man, chimpanzee and perhaps elephants recognize themselves in a mirror, but neither do the other primates nor any other animal. Thus, the self-consciousness of dogs is probably less wide than the self-consciousness of men.

Dog is sensitive to reward and punishment; it recognizes the emotions of other dogs and perhaps of men. So, like men, the dog is able to inhibit the search of immediate pleasure by the fear of punishment; this is a mark of self-control. Must we conclude that dogs have some moral sense?

REFERENCES

[1]   Berlyne DE. The arousal and satiation of perceptual curiosity in the rat. J Comp Physiol Psychol 1955; 48(4) 238-246
[2]   Darwin C. The Expression of the Emotions in Man and Animals. John Murray Ed. London 1872.
[3]   Kalisher O: Das Grosshirn des Papageien in anatomischer und physiologishe Beziehung. Physik. Abh. Kgl. Preuss. Ak. Wissench. ed. 1905; vol 4 : p. 104.
[4]   Oriol C. Compagnon de Safari. Guide pratique de la faune namibienne. Chez l'auteur Ed. Windhoek, Namibia 2003; pp.116. In French.
[5]   Chomsky N. Current Issues in Linguistic Theory. Mouton ed., The Hague 1964.
[6]   Penfield W, Rasmussen T. The cerebral cortex of man; a clinical study of localization of function. Macmillan ed.. Oxford, England 1950; p. 248.
[7]   Bechara A, Damasio H, Damasio AR. Emotion, decision making and the orbitofrontal cortex. Cereb Cortex 2000; 10(3): 295-307. Review.

[8]   Gazzaniga MS, Ledoux JE, Wilson DH. Language, praxis, and the right hemisphere: clues to some mechanisms of consciousness. Neurology 1977; 27(12): 1144-7.

[9]   Crick F. The Astonishing Hypothesis: The Scientific Search for the Soul. Scribner Ed. New-York NJ USA1990; p. 336.

[10]   Baars BJ. In the Theater of Consciousness: The Workspace of the Mind. NY: Oxford University Press 1997.

[11]   Edelman GM, Toroni G. A universe of Consciousness: How Matter Becomes Imagination., Basic Books ed. New-York NJ, USA 2000.

[12]   Dehaene S, Naccache L. Neural signature of the conscious processing of auditory regularities. Proc Natl Acad Sci USA 2009; 106 (5) 1672-7.

[13]   Naccache L, Dehaene S, Cohen L, *et al.* Effortless control: executive attention and conscious feeling of mental effort are dissociable. Neuropsychologia 2005; 43(9): 1318-28.

[14]   Bechara A, Damasio H, Damasio AR. Emotion, decision making and the orbitofrontal cortex. Cereb Cortex 2000; 10(3): 295-307. Review.

[15]   Freedman M, Black S, Ebert P, Binns M. Orbitofrontal function, object alternation and perseveration. Cereb Cortex 1998; 8(1): 18-27.

Additional Information
NOTATIONS AND USEFUL CONSTANTS

Main Notations

c_K, c_{Na}:	Ionic concentrations
f, ω:	Damping and pulsation of little oscillations
F:	Firing rate
γT:	Linearity index of synaptic coupling
i:	Electric current density
I:	Electric current
n_R:	Proportion of operational receptors in a conditional synapse
R:	Radius of the triggering area
t:	Time
T:	Time constant
T in kT:	Temperature
$U = V - V_R$; U_{Thr}:	Threshold
V:	Cell electric potential
V_R:	Rest potential
V_K, V_{Na}:	Nernst's potentials
W:	Energy
Ω:	Proportion of open post-synaptic pores

Some Useful Constants

1 mole = 6.025 10^{23} ions

electron charge: e = 1.6 10^{-19} Coulombs electron-volt: 1 eV = 1.6 10^{-19} J

Boltzmann constant: k = 1.38 10^{-16} cgs Planck constant: h = 6.62 10^{-27} cgs

T = 273+37 = 310 d°K kT/e = 25 mV W (Kcal/mole) = 0.62 [w/kT]

$\hbar$ = h / ($2\pi kT$) = 2.46 10^{-12} s

Index

www.ingramcontent.com/pod-product-compliance
Lightning Source LLC
Chambersburg PA
CBHW050811220326
41598CB00006B/180